2nd Workshop on Germanium Detectors and Technologies 2014

AA009058

Journal of Physics: Conference Series Volume 606

Vermillion, South Dakota, USA
14-17 September 2014

Editors:

Iris Abt
Christina Keller
Guojian Wang
Bela Majorovits
Dongming Mei
Wenzhao Wei

ISBN: 978-1-5108-0358-9
ISSN: 1742-6588

Printed from e-media with permission by:

Curran Associates, Inc.
57 Morehouse Lane
Red Hook, NY 12571

Some format issues inherent in the e-media version may also appear in this print version.

Printed by Curran Associates, Inc. (2015)

For permission requests, please contact the Institute of Physics
at the address below.

Institute of Physics
Dirac House, Temple Back
Bristol BS1 6BE UK

Phone: 44 1 17 929 7481
Fax: 44 1 17 920 0979

techtracking@iop.org

Additional copies of this publication are available from:

Curran Associates, Inc.
57 Morehouse Lane
Red Hook, NY 12571 USA
Phone: 845-758-0400
Fax: 845-758-2634
Email: curran@proceedings.com
Web: www.proceedings.com

TABLE OF CONTENTS

DARK MATTER

NEUTRINOLESS DOUBLE BETA DECAY ($0\nu\beta\beta$)

NEUTRINO PHYSICS

GEMANIUM CRYSTAL GROWTH

FACILITES AND APPLICATIONS

Journal of Physics: Conference Series **606** (2015) 011001 doi:10.1088/1742-6596/606/1/011001

Editorial Committee: Iris Abt, Béla Majorovits, Christina Keller, Dongming Mei, Guojian Wang, Wenzhao Wei

(1) Preface

The 2nd workshop on Germanium (Ge) detectors and technology was held at the University of South Dakota on September 14-17th 2014, with more than 113 participants from 8 counties, 22 institutions, 15 national laboratories, and 8 companies. The participants represented the following big projects: (1) GERDA and Majorana for the search of neutrinoless double-beta decay (0νββ); (2) SuperCDMS, EDELWEISS, CDEX, and CoGeNT for the search of dark matter; (3) TEXONO for sub-keV neutrino physics; (4) AGATA and GRETINA for gamma tracking; (5) AARM and others for low background radiation counting; (5) as well as PNNL and LBNL for applications of Ge detectors in homeland security. All participants have expressed a strong desire on having better understanding of Ge detector performance and advancing Ge technology for large-scale applications.

The purpose of this workshop was to leverage the unique aspects of the underground laboratories in the world and the germanium (Ge) crystal growing infrastructure at the University of South Dakota (USD) by bringing researchers from several institutions taking part in the **Experimental Program to Stimulate Competitive Research** (EPSCoR) together with key leaders from international laboratories and prestigious universities, working on the forefront of the intensity to advance underground physics focusing on the searches for dark matter, neutrinoless double-beta decay (0νββ), and neutrino properties. The goal of the workshop was to develop opportunities for EPSCoR institutions to play key roles in the planned world-class research experiments. The workshop was to integrate individual talents and existing research capabilities, from multiple disciplines and multiple institutions, to develop research collaborations, which includes EPSCoR institutions from South Dakota, North Dakota, Alabama, Iowa, and South Carolina to support multi-ton scale experiments for future.

The topic areas covered in the workshop were: 1) science related to Ge-based detectors and technology; 2) Ge zone refining and crystal growth; 3) Ge detector development; 4) Ge orientated business and applications; 5) Ge recycling and recovery; 6) introduction to underground sciences for young scientists; and 7) introduction of experimental techniques for low background experiments to young scientists. Sections 1-5 were dedicated to Ge detectors and technology. Each topic was complemented with a panel discussion on challenges, critical measures, and R&D activities. Sections 6-7 provided students and postdocs an opportunity to understand fundamental principles of underground sciences and experimental techniques on low background experiments. To these two sections, well-known scientists in the field were invited to give lectures and allow young scientists to make presentations on their own research activities.

Fifty-six invited talks were delivered during the three-day workshop. Many critical questions were addressed not only in the specific talks but also in the panel discussions. In addition to various progress related to germanium detectors and technology that were discussed in the presentations from the respective experiments, the following panel questions were discussed in depth:

Panel Section 1: Panel discussion on germanium crystal growth
a) Type, size, quality, and quantity demands on crystal growth. How do the available suppliers work together to deliver?
b) Characterization of electric and optical properties of crystals. How do the available suppliers share information and cross exam?
c) Crystal orientation versus impurity and dislocation distribution. How can (100), (111) and other axes orientations suit the different physics goals?

Panel Section 2: Panel discuss on detector development
a) Low-energy threshold techniques. How do contacts, front-end electronics, and cryostat impact the threshold?
b) Improving timing response. How can new contacts help to eliminate transition layer effects and hence slow pulses?
c) Passivation and surface contamination. How can surface contamination be reduced by controlling processing, etching, and coating?

Panel Section 3: Surface contamination and the standardized assay program
a) Radon daughter deposition on the surface of germanium and metals used for contacts. How to remove radon daughters from the surface - Ethylenediaminetetraacetic acid (EDTA) detergent and other solutions.
b) Low-level ^{210}Pb measurements. How to effectively measure alphas, betas and gamma rays from ^{210}Pb decays to achieve sensitivities below 1 mBq/m^2?
c) Assay procedure and reporting standards. How to standardize assay procedures and reporting algorithms to reduce systematic uncertainties in ultra low level counting?
d) AARM and PNNL assay consortium discussion.

Nineteen papers published in these proceedings represent invited talks given in the workshop for germanium detectors and technology related science – dark matter, neutrinoless double-beta decay, low energy neutrino physics, gamma tracking, and high-purity germanium crystal growth as well as germanium detectors application in low-background counting, homeland security, and underground laboratory facility.

A global collaboration, including experts and unique facilities from six countries, on advanced germanium detectors and technology was established as the workshop participants recognized this an important need to advance the technology. The vision of this collaboration is to develop an international partnership by combining individual talents and existing research capabilities from nine institutions. The focus is on development of new Ge detector technologies necessary for ton-scale germanium experiments. This could have impact on direct dark matter detection, the search for $0\nu\beta\beta$ decay, as well as neutrino-nucleus coherent scattering (νN-coh), probing yet-unknown neutrino properties like neutrino magnetic moments (μ_n) and milli-charge ($\delta_Q[\nu]$) while advancing other applications.

Conferences Photos

(2) Sponsor or funding acknowledgements
National Science Foundation, US, Award number: NSF-IIA 1434142.
South Dakota Governor's Office Economic Development
South Dakota Board of Regents
South Dakota Small Business Innovation Research
Vermillion Area Chamber of Commerce & Development Company
Skysprings Nano materials, Inc., Houston, TX
The University of South Dakota

Journal of Physics: Conference Series **606** (2015) 011001 doi:10.1088/1742-6596/606/1/011001

(3) List of committees

Table 1: Proposed Steering Committee: (70% are from the EPSCoR jurisdictions with 18% women)

Name	Role	Affiliation
Mary Berry	Chair	The University of South Dakota, US
James Rice	Member	South Dakota State University, US
Paul Turman	Member	South Dakota Board of Regents, US
Mel Ustad	Member	Governor's Office Economic Development, SD, US
Christina Keller	Member	The University of South Dakota, US
Steve Elliott	Member	Los Alamos National Laboratory, US
Kevin Lesko	Member	Lawrence Berkeley National Laboratory, US
Blas Cabrera	Member	Stanford University, US
Jason Detwiler	Member	University of Washington, US
Bernard Sadoulet	Member	UC Berkeley, US
Bela Majorovits	Member	Max-Planck-Institute for Physics, Germany
Dongming Mei	Member	The University of South Dakota, US
Tom Lograsso	Member	Ames National Laboratory, US
Nuri Oncel	Member	University of North Dakota, US
Frank Avignone	Member	University of South Carolina, US
Wayne Barkhouse	Member	University of North Dakota, US
Vince Guiseppe	Member	University of South Carolina, US
Jerry Busenitz	Member	University of Alabama, US
Andreas Piepke	Member	University of Alabama, US
Mike Headley	Member	Sanford Underground Research Facility, US
Barbara Szczerbinska	Member	Dakota State University, US
Kara Keeter	Member	Black Hills State University, US

Table 2: Local Organization Committee: (100% are from the EPSCoR jurisdiction with 40 % women)

Name	Role	Affiliation
Dongming Mei	Chair	The University of South Dakota, US
Christina Keller	Member	The University of South Dakota, US
Joel Sander	Member	The University of South Dakota, US
Ryan Martin	Member	The University of South Dakota, US
Mary Nielsen	Member	The University of South Dakota, US
Wenzhao Wei	Member	The University of South Dakota, US

Table 3: Program Committee: (55% are from the EPSCoR jurisdictions and17% are female scientists)

Name	Role	Affiliation
Dongming Mei	Chair	The University of South Dakota, US
Bela Majorovites	Member	Max-Planck-Institute for Physics, Germany
Xavier-Francois Navick	Member	GEA, Centre d'Etudes Saclay, IRFU, France
Henry Wong	Member	Institute of Physics, Academic Sinica, Taiwan
Jianmin Li	Member	Tsinghua University, Beijing, China
Steve Elliott	Member	Los Alamos National Laboratory, US
Blas Cabrera	Member	Stanford University, US
John Simpson	Member	CCLRC Daresbury Laboratory, UK
Prisca Cushman	Member	University of Minnesota, US
Adam Bernstein	Member	Lawrence Livemore National Laboratory, US
Yuen-Dat Chan	Member	Lawrence Berkeley National Laboratory, US

Ryan Martin	Member	The University of South Dakota, US
Joel Sander	Member	The University of South Dakota, US
Kara Keeter	Member	Black Hills State University, US
Barbara Szczerbinska	Member	Dakota State University, US
Stanley Howard	Member	South Dakota School of Mines and Technology, US
Tom Lograsso	Member	Ames National Laboratory, US
Nuri Oncel	Member	University of North Dakota, US
Frank Avignone	Member	University of South Carolina, US
Wayne Barkhouse	Member	University of Alabama, US
Vince Guiseppe	Member	University of South Carolina, US
Jerry Busenitz	Member	University of Alabama, US
Andreas Piepke	Member	University of Alabama, US
Brianna Mount	Member	Black State University, US

4) The workshop website

The detailed information and all 56 talks can be found at:
http://www.geworkshop.org/indico/conferenceDisplay.py?confId=0.

Peer review statement

All papers published in this volume of *Journal of Physics: Conference Series* have been peer reviewed through processes administered by the proceedings Editors. Reviews were conducted by expert referees to the professional and scientific standards expected of a proceedings journal published by IOP Publishing.

Status and prospects of the China Dark Matter Experiment

Shukui Liu and Qian Yue[a]

Department of Engineering Physics, Tsinghua University, Beijing, 100084, China
E-mail: [a] yueq@mail.tsinghua.edu.cn

Abstract. The China Dark Matter Experiment (CDEX) pursues direct searches of light Weakly Interacting Massive Particles (WIMPs) at the China Jinping Underground Laboratory (CJPL), which is the deepest operating laboratory for astroparticle research in the world. Results from a prototype CDEX-0 20 g germanium detector array and CDEX-1 994 g *p*PCGe (p-type Point Contact Germanium) detector are reported. The new result from CDEX-1 *p*PCGe excludes the CoGeNT-2013 allowed region with an identical detector technique. CDEX-10 with a PCGe array of 10 kg target mass range enclosed in a liquid argon anti-Compton detector is being constructed and tested. The CDEX program evolves into the targets of "CDEX-1T Experiment". The CDEX-1T experiment will be located in CJPL-II which is under construction and will be finished by the end of 2016.

1. Introduction

There is compelling evidence that about one-quarter of the energy density of the Universe is made up of Dark Matter, the identification and study of which are among the most important goals in basic research [1]. WIMPs is a most favoured candidate for Dark Matter. Some earlier experiments have produced intriguing results which can be interpreted as positive signatures of WIMPs with relatively light mass of a few GeV[2-5], in disagreement to other experiments which yielded negative results[6-10]. The issue remains controversial and generates intense interest.

The CDEX [11] pursues formally the direct detection of light WIMPs from 2009 on, after more than five-year pre-study at Tsinghua University, towards the goal of a ton-scale germanium detector array. As illustrated schematically in Figure 1, the CDEX experiment is performed in the China Jinping Underground Laboratory (CJPL) [12] which is the deepest operating laboratory for astroparticle research in the world, with more than 2400m of rock overburden, attenuating the cosmic-rays background by more than a factor of 10^8 to the level of about 60 m^{-2} y^{-1} [13]

Figure 1. Schematic diagram of the CDEX-1 experiment conducted in the China Jinping Underground Laboratory

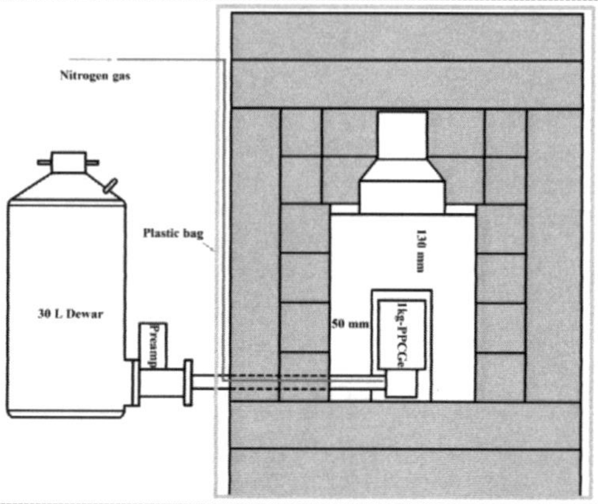

Figure 2. Schematic diagram of the CDEX-1 experiment, including PCGe and NaI(Tl) detectors and copper shielding.

The CDEX experiment is led by Tsinghua University, with collaborating partners from Sichuan University, Nankai University, China Institute of Atomic Energy, Yalong River Hydropower Development Company, as well as the TEXONO Collaboration. As the phase-I experiment by CDEX collaboration, CDEX-1 deploys state-of-the-art ultra-low-background point-contact 1kg-mass germanium detectors with sub-keV sensitivities [14]. This germanium detector is enclosed by NaI(Tl) crystal scintillator plus copper blocks and placed inside a poly-ethylene housing, as shown in figure 2. These can highly suppress the gamma background and neutron radiations from outside. In addition, the inner space of the copper is purged by nitrogen gas to reduce the influence of the radioactive radon gas.

2. The current status of CDEX

Ultra-low energy threshold germanium detectors were identified [15] as effective means to probe the light WIMPs and motivated by the development of point-contact germanium detectors. The generic "benchmark" goals in terms of detector performance are: (1) modular target mass of order of 1 kg and total mass up to 10kg; (2) detector sensitivities reaching to the range of 100 eV; (3) background at the range of 1 kg^{-1} keV^{-1} day^{-1} (cpkkd).

Results on light WIMPs from the CDEX-0 and CDEX-1 employed a germanium detector array with a total mass of 20 g and a crystal mass of 994g *p*PCGe detector respectively had been published [16,17]. The latest results from CDEX-1 experiment with the 994g *p*PCGe detector having two crucial features incorporated including the presence of an anti-Compton detector and surface event suppression were recently reported and shown in Figure 3 [18]. All events above the analysis threshold of 475 eVee can be quantitatively accounted for with the understood background channels. An order of magnitude improvement over the previous CDEX results is achieved. In particular, the CoGeNT-2013 allowed region is excluded based on identical detector technique. It provides direct comparisons without the uncertainties due to model dependence introduced via the choice of astrophysical and nuclear parameters.

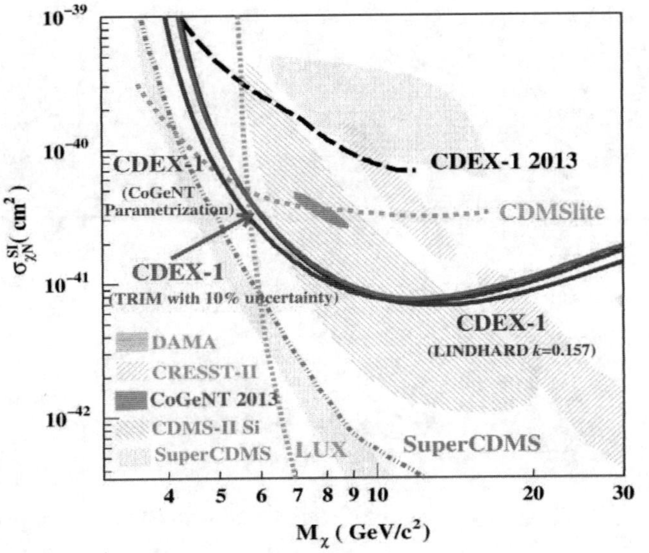

Figure 3 Exclusion plot of WIMP-nucleon cross-section versus WIMP mass, showing the latest CDEX-1 results and the previous positive regions

3. Prospects of CDEX

The CDEX-1 experiment continues to accumulate data at CJPL. Research programs are pursued to further reduce the physics threshold via hardware and software efforts. Time modulation of the data will be studied. The CDEX-10 experiment with a PCGe array of 10 kg target mass range enclosed in an active liquid argon anti-Compton detector is being constructed. Feasibility studies towards scale-up to ton-scale experiment are being pursued.

The CDEX program evolves into the targets of "CDEX-1T Experiment" with ton-scale germanium detector arrays, which will compose of thousands of kg-mass prototype germanium detectors. The detector arrays will be submerged in liquid argon for cooling and active shielding. The conceptual drawing is shown in Figure 4. This Facility will further contribute to the studies of Dark Matter and Neutrinoless Double Beta Decay. It will be located in one of the four halls with dimensions of 14m(W)×14m(H)×65(L) in CJPL-II which will be completed by the end of 2016.

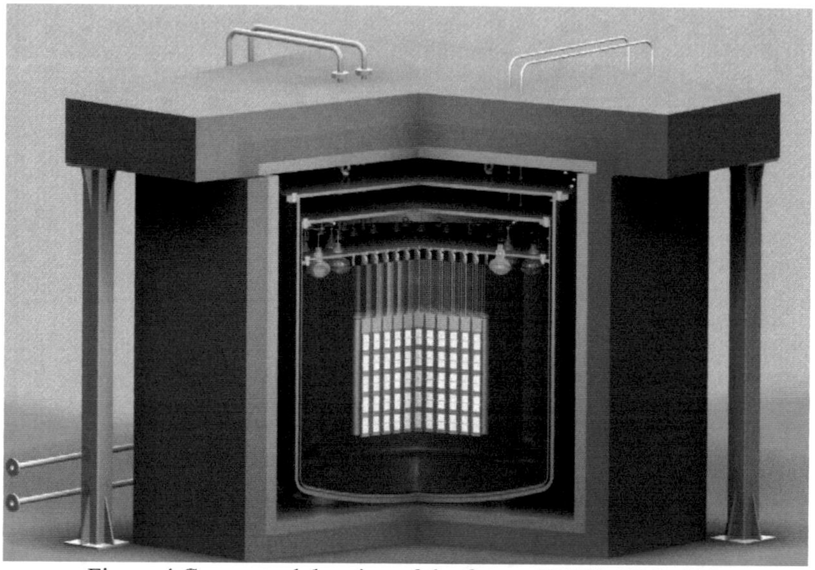

Figure 4 Conceptual drawing of the future CDEX-1T facility

References

[1] M. Drees and G. Gerbier, Phys. Rev. D 86, 289 (2012), and references therein
[2] R. Bernabei et al., Eur. Phys. J. C 56, 333 (2008); 67, 39 (2010).
[3] C.E. Aalseth et al., Phys. Rev. D 88, 012002 (2013); arXiv:1401.3295.
[4] G. Angloher et al., Eur. Phys. J. C 72, 1971 (2012).
[5] R. Agnese et al., Phys. Rev. Lett. 111, 251301 (2013).
[6] E. Aprile et al., Phys. Rev. Lett. 109, 181301 (2012).
[7] D. S. Akerib et al., Phys. Rev. Lett. 112, 091303 (2014).
[8] H. B. Li et al., Phys. Rev. Lett. 110, 261301 (2013).
[9] R. Agnese et al., Phys. Rev. Lett. 112, 041302 (2014).
[10] R. Agnese et al., Phys. Rev. Lett. 112, 241302 (2014).
[11] K.J. Kang et al., Front. Phys. 8, 412 (2013)
[12] K.J. Kang et al., J. Phys. Conf. Ser. 203, 012028 (2010)
[13] Y.C. Wu et al., Chin. Phys. C 37, 086001 (2013)
[14] P.N. Luke et al, IEEE Trans Nucl. Sci. 36, 926 (1989);
[15] K.J. Kang et al., Chin. Phys. C 37 , 126002 (2013).
[16] S.K. Liu et al., Phys. Rev. D 90, 032003 (2014).
[17] W. Zhao et al. Phys. Rev. D 88, 052004 (2013);
[18] Q. Yue et al., Phys. Rev. D 90, 091701R (2104)

Status of the EDELWEISS-III Dark Matter search

Cécile Kéfélian

Karlsruhe Institute of Technology, Institut für Kernphysik, Postfach 3640, 76021 Karlsruhe, Germany

Institut de Physique Nuclaire de Lyon, Campus La Doua, rue Enrico Fermi, 69622 Villeurbanne, France

E-mail: cecile.kefelian@kit.edu

Abstract. EDELWEISS is a direct dark matter search program looking for Weakly Interacting Massive Particles (WIMPs) in the GeV-TeV mass range using an array of cryogenic Ge mono-crystals. These high-performance detectors are read out simultaneously by NTD thermal sensors and by surface electrodes. They are installed in the deepest European underground laboratory in Modane. The third phase of the experiment is currently ongoing with a major upgrade of the setup. New FID800 Ge bolometers have been developed and the shielding has been improved to lower the background. In addition, the cryogenic and acquisition systems have been upgraded to improve energy resolutions and thresholds.

1. The EDELWEISS experiment

1.1. Experimental setup

The EDELWEISS experiment is operated in the deepest underground laboratory in Europe, the *Laboratoire Souterrain de Modane* (LSM). The 4800 m.w.e mountain rock thickness above the experiment allows to reduce the cosmic muon flux by a factor 10^6, down to 5 muons/m^2/day [1]. Remaining muons going through the experiment are tagged using an active muon veto system surrounding the setup (Fig. 1 left). Due to the 46 modules of plastic scintillator covering a surface of 100 m^2, a geometric coverage of 98% for through-going muons is achieved. In addition, a 50 cm layer of polyethylene (PE) shield is used to moderate radiogenic neutrons. A 20 cm lead shield is used to stop external γ and β, which are further reduced in the copper of the cryostat where the Ge detectors are cooled down to 18 mK. Inside the cryostat, additional pieces of lead and PE are used to shield the bolometers from the cold electronics (Fig. 1 right). Thanks to the external shields, the background in the detectors is mainly coming from inside the cryostat.

1.2. Germanium bolometers

The EDELWEISS detectors are high-purity Ge mono-crystal bolometers equipped with Neutron Transmutation Doped Ge sensors used as thermistors and covered by concentric aluminum ring electrodes. Due to the simultaneous measurement of heat and ionization, it is possible to distinguish on an event-by-event basis electronic recoils (ER) induced by γ or β from nuclear recoils (NR) induced by WIMPs or neutrons. Indeed, the ratio of ionization over recoil energy, the so-called ionization yield Q, is equal to 1 for ER by normalization to ^{133}Ba calibration and $\simeq 0.3$ for NR as described by the Lindhard formula [2]. This powerful discrimination is

Journal of Physics: Conference Series **606** (2015) 012002 doi:10.1088/1742-6596/606/1/012002

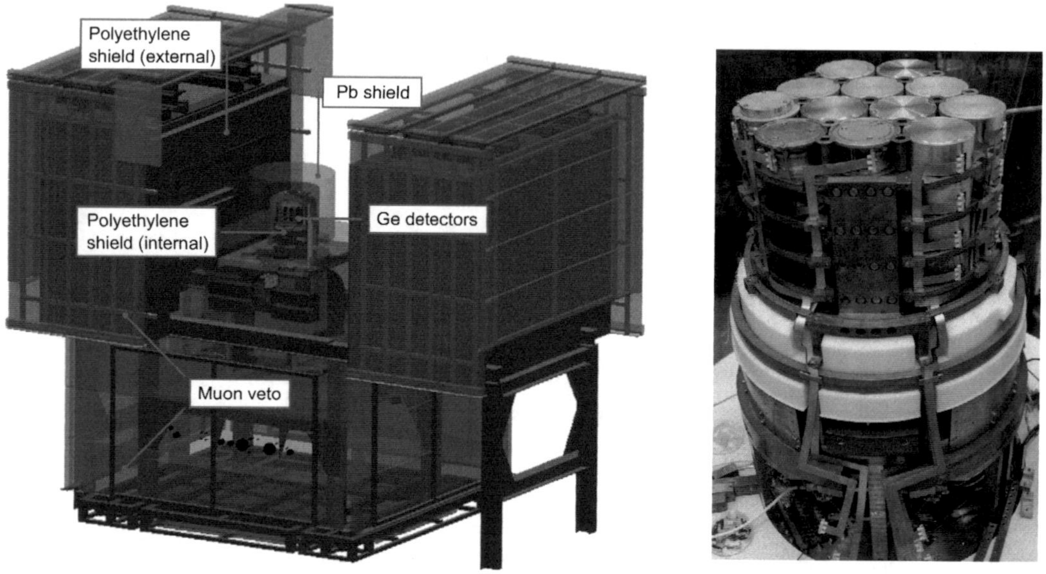

Figure 1. Left: schematic of the EDELWEISS-III setup showing in the center the cryostat hosting the bolometers, surrounded by passive lead and polyethylene (PE) shields and an active muon veto in order to protect the detectors from various backgrounds. Right: view of the opened cryostat equipped with 36 FID germanium bolometers in their copper housing, which are shielded from the cold electronics below using internal lead and PE shields.

reinforced by the rejection of surface events which are most likely α and β particles. These surface events are particularly dangerous as they may suffer from bad charge collection and are thus susceptible to leak into the NR band where WIMP signals are expected. The discrimination of such events is possible thanks to the electrode design: electrodes are alternatively biased ($O(1\,\mathrm{V})$) and bounded, defining a fiducial volume for which the charge collection is complete and a surface volume for which it may not. One set of electrodes, the collecting electrodes, has a higher bias and collects charges from the bulk whereas events with a signal on the other set (veto electrodes) are efficiently rejected. Indeed, as can be seen Fig. 2, the electric field lines in the fiducial volume are perpendicular to the detector surface and charges are drifted to the collecting electrodes at the top and the bottom of the detector. At the surface, the field lines are parallel to the detector surface and charges are drifted to the veto and collecting electrodes on one side. Thus, by rejecting events with veto signal, surface events can be discriminated.

2. Results from the second phase of the experiment EDELWEISS-II

Before detailing the numerous improvements which were carried out for EDELWEISS-III, here is a review of EDELWEISS-II results and what we learned from it. EDELWEISS-II was originally designed to detect WIMPs in the mass range [10,100 GeV]. For this purpose, 10 so-called *Inter-Digitized* (ID) detectors of 400 g each were run for a continuous data taking period of 13 months from April 2009 to May 2010.

2.1. Standard WIMP analysis

The standard WIMP analysis [4] is based on a total effective exposure of 384 kg·days after all cuts. 5 events were observed in the 90% C.L. NR band in the region of interest with recoil

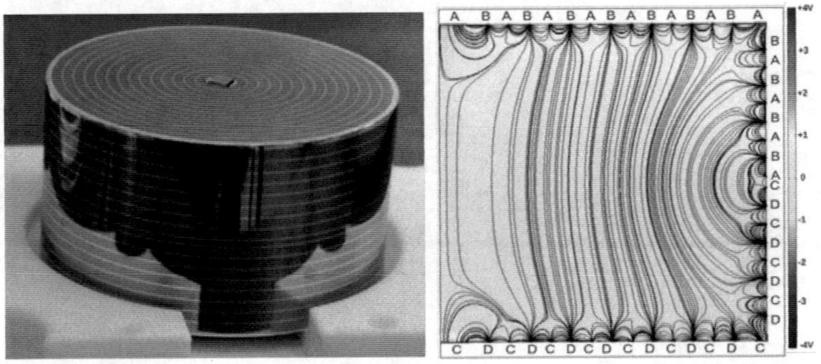

Figure 2. Photo and sketch of the electric field lines for a cross section of a FID800 detector. The color code indicates the electric potential, the electric field lines being drawn in black. Charges induced by fiducial events are collected by the B and D electrodes only whereas charges induced by surface events are partly collected by at least one of the veto electrodes A and C.

energy [20, 200 keV] whereas 3.0 events were expected from background (1.4 neutrons from ambient radioactivity, 0.9 γ-ray leaking, 0.4 muon-induced neutrons and 0.3 unrejected surface events). These 5 events were interpreted as an upper fluctuation of the background and limits on the spin-independent (SI) WIMP-nucleon cross section were derived, leading to an exclusion limit of $\sigma_{SI} < 4.4 \times 10^{-8}$ pb at 90% C.L. for a WIMP mass of 85 GeV (Fig. 3 left). Limits on spin-dependent interaction cross-section were also derived [4]. As both CDMS and EDELWEISS experiments use the same target, their results were combined [5] leading to an exclusion limit of $\sigma_{SI} < 3.3 \times 10^{-8}$ pb (90% C.L.) for a WIMP mass of 90 GeV, based on a combined exposure of 614 kg·days.

2.2. Low mass WIMP analysis

In order to investigate the various signal claims at low mass WIMPs, a restricted data set from EDELWEISS-II was used by selecting detectors with low background and good performance at low energies down to 5 keV. By choosing an upper threshold of 20 keV, this analysis is completely independent from the standard WIMP analysis and covers a WIMP mass range of [7, 30 GeV]. Four out of the 10 ID detectors were used and stronger quality cuts were applied, leading to an exposure of 113 kg·days. Up to 3 events (depending on the WIMP mass) were observed in the WIMP search region, compatible with the expected background of 2.9 events (< 1.7 ambient neutrons and < 1.2 misidentified γ's). From this measurement, an exclusion limit of $\sigma_{SI} < 1.0 \times 10^{-5}$ pb at 90% C.L. was derived (Fig. 3 left), which was excluding DAMA/LIBRA [7] and CRESST [8] claims and significantly constraining the CoGeNT [9] signal.

2.3. Axion search

Axions are hypothetical dark matter particles which could induce an electronic recoil in the bolometers after production of γ's via the Primakoff effect or electrons via the axio-electric effect. The EDELWEISS-II sensitivity to electronic recoils down to 2.5 keV makes the experiment sensitive to axions. Due to the low-radioactivity setup and the fiducial volume selection, a background level down to 0.3 events/kg/day/keV could be achieved. The full EDELWEISS-II data set with relaxed cuts was used, with an exposure of 484 kg·days. Axion production in the Sun and axions as dark matter particles were studied considering different scenarios on the axion couplings with photons and electrons, and the effective axion-nucleon coupling for ^{57}Fe [10]. Combining all obtained results, one can exclude the mass range 0.91 eV < mA < 80 keV

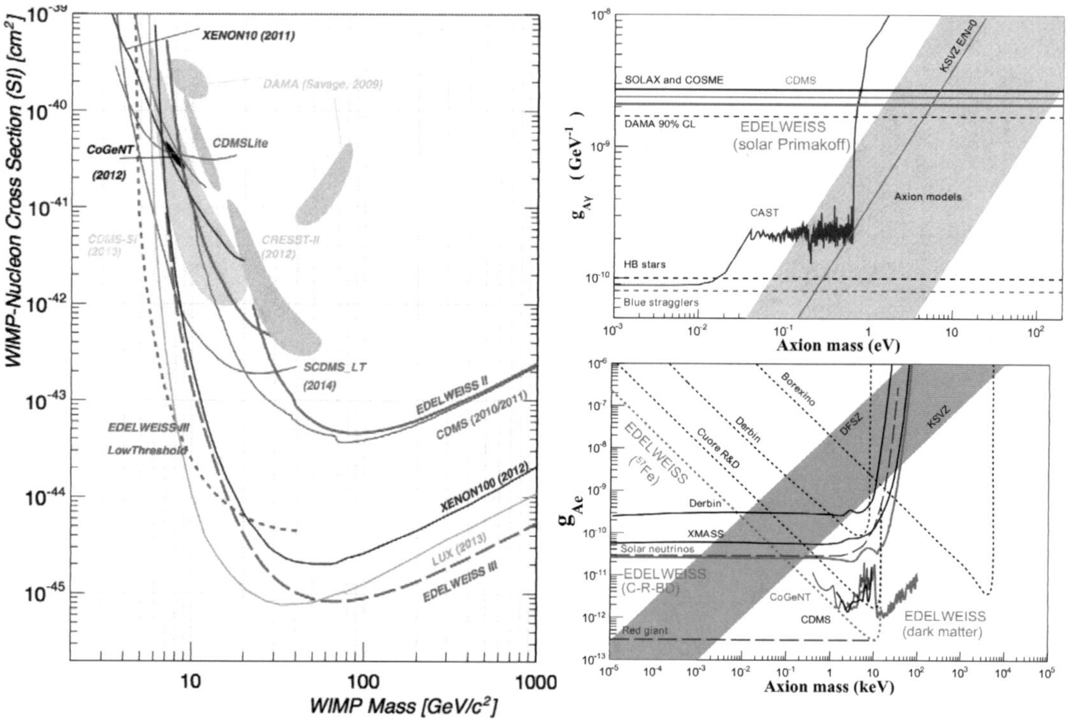

Figure 3. Left: Upper limits on the spin-independent WIMP-nucleon interaction cross section versus the WIMP mass. EDELWEISS-II results corresponding to [4] and [6] are drawn with solid red lines. EDELWEISS-III projection for standard and low mass WIMPs are marked using red dotted lines. Right: EDELWEISS-II limits on $g_{A\gamma}$ coupling from the solar Primakoff flux and on g_{Ae} coupling versus the axion mass (red lines) [10].

for Dine-Fischler-Srednicki-Zhitnitskii (DFSZ) axions and 5.73 eV < mA < 40 keV for Kim-Shifman-Vainstein-Zakharov (KSVZ) axions (Fig. 3 right).

3. Improvements for EDELWEISS-III

The EDELWEISS-II sensitivity reached in 2010 was mainly limited by misidentified γ's and radiogenic neutrons produced by (α,n) reactions in the cabling inside the cryostat. In addition, the low mass WIMP analysis showed the necessity to improve energy resolutions and lower the thresholds. In order to achieve the EDELWEISS-III goal to probe WIMP-nucleon cross sections down to 10^{-9} pb, many upgrades were made to reduce the backgrounds and improve performance at low energies. 36 so-called *Fully Inter-Digitized* (FID) detectors of a mass of 800 g each with improved background rejection capabilities were installed in the cryostat. These new detectors are twice more massive than the previous generation and benefit from a fiducial volume increase from 40% to 75% of the detector volume. This was achieved thanks to the new electrode design which replaced the planar electrodes on the lateral surface with ring electrodes. Thus the detection mass was increased from 1.6 kg in EDELWEISS-II to more than 20 kg in EDELWEISS-III.

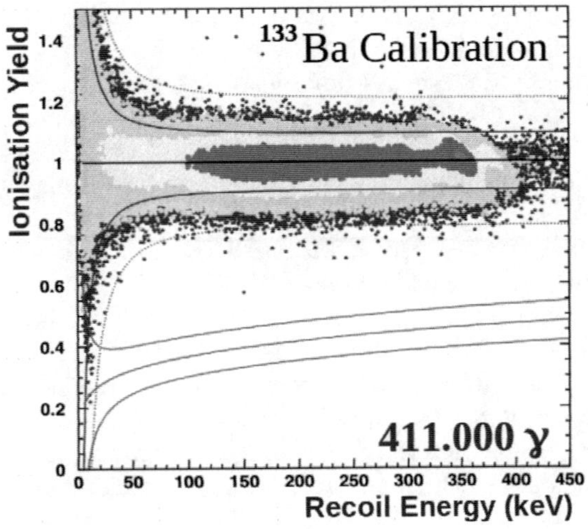

Figure 4. Ionization yield as a function of the recoil energy in keV for events from γ-calibration using a ^{133}Ba source. The dashed and continuous blue lines are respectively the 90% C.L. and the 99.98% C.L. electronic recoil bands whereas the continuous red lines corresponds to the 90% C.L. nuclear recoil band. There is no single γ event leaking in the nuclear recoil band over the 411 000 measured γ.

3.1. Passive background rejection

In order to reduce the neutron background by a factor 100 down to $(0.8 - 1.9) \times 10^{-4}$ events/kg·day,cables inside the cryostat were replaced by designed in-house radiopure Cu Kapton cabling. Additional internal and external PE shields were installed, notably between the internal roman lead shield and the bolometers to protect against neutrons from the cold electronics mainly arising from (α,n) reactions (Fig.1 right). One also expects a decrease of the internal γ-background by a factor between 2 and 6 down to 14-44 events/kg·day thanks to the new low activity (NOSV) copper screens of the cryostat. It should be noted that the radon level near the bolometers is controlled and reduced down to a few tens of mBq/m^3. Efforts are also ongoing to reduce the muon-induced neutron background by at least a factor 10. For this purpose, 4 additional scintillator panels were installed to cover the detection gap located above the cryostat. A calibration campaign is currently ongoing to determine the position-dependent module response which will, together with a detailed MC simulation of the muon propagation, enable a more precise determination of the muon-induced neutron background for the modified EDELWEISS-III setup.

3.2. Active background rejection

The active rejection of misidentified γ's which were limiting the sensitivity in the previous phase was shown to be considerably improved in EDELWEISS-III. This background was arising from multiple γ-scattering which deposit energy in the bulk and near the detector surface but with a veto signal below threshold. These events can now be rejected more efficiently due to the new electrode design. The rejection power of γ's was measured using a ^{133}Ba source and determined to be more than 5 times better than in EDELWEISS-II: out of the $> 4 \times 10^5$ measured γ's, none leaked into the NR band above 20 keV, leading to a rejection factor $< 6 \times 10^{-6}$ NRs/γ (Fig.4). The new electrode configuration also induces a better discrimination power of surface events which was measured using a ^{210}Pb source to be >1.5 times better than in EDELWEISS-II with $< 4 \times 10^{-5}$ misidentified β's per evt/kg·day in EDELWEISS-III [11].

3.3. Energy resolutions and threshold

New cryogenic and electronic systems were installed to improve the performance at low energies. In order to reduce the microphonic noise that decreases the heat resolution, the pulse tubes near the cryostat were replaced by Gifford-McMahon thermal machines outside of the shields, connected to the cryostat using a cryoline. Concerning the electronic system, the active feedback system necessary to reset the bolometers' ADC was replaced by a relay system. No more needed resistors were removed to avoid Johnson noise and thus improve the noise level at low energies. These upgrades lead to the lowering of the average FWHM of the heat baselines from 1.2 keV to 1.0 keV and of the ionization baselines from 900 eV to 600 eV. Thanks to these improvements, a lower analysis threshold can be achieved as well as a sensitivity down to 5 keV necessary for the detection of low mass WIMPs. In addition, the installation of an integrated DAQ system improves the reliability of the data taking and the event building: all the individual bolometer data are processed through a single crate which provides a common clock for the bolometer data acquisition as well as for the muon veto system. In addition, the DAQ system is able to do a real time processing of the data stream (e.g. take a trigger decision) in parallel for all channels. Based on this fast trigger decision, it also allows to read out fast sampled (40MS/s) traces of ionization channels which provide time-resolved ionization signals. This additional information can be used to localize the event vertex along the z-axis or identify multiple scattering events within the detector. We are currently studying the potential of this new data to reject multiple scattering events as well as to have complementary tool for surface event rejection.

4. Conclusion

EDELWEISS-III is currently taking data since August 2014, aiming at first to obtain a background-free exposure of 3000 kg·days at the end of 2015. The final goal is to measure 12000 kg·days exposure for which background should start to appear, leading to a sensitivity to a spin-independent WIMP-nucleon cross section of 10^{-9} pb.

Acknowledgments

The help of the technical staff of the Laboratoire Souterrain de Modane and the participant laboratories is gratefully acknowledged. The EDELWEISS project is supported in part by the Helmholtz Association, by the French Agence Nationale pour la Recherche, by Science and Technology Facilities Council (UK) and the Russian Foundation for Basic Research (grant No. 07-02-00355-a). R&D activities towards EURECA are supported in part by the German ministry of science and education (BMBF Verbundforschung ATP Proj.-Nr. 05A11VK2) and by the Helmholtz Alliance for Astroparticle Physics (HAP), funded by the Initiative and Networking, Fund of the Helmholtz Association. In addition, C.K. acknowledges support from the DFG-funded graduate school KSETA: Karlsruhe School of Elementary Particle and Astroparticle Physics: Science and Technology.

References

[1] Schmidt B *et al.* 2013 *Astrop. Phys.* *44* 28
[2] Lindhard J *et al.* 1963 *Mater. Phys. Med. Dan. Vid. Selsk* 10 1
[3] Broniatowski A *et al.* 2009 *Phys. Lett.* **B681** 305
[4] Armengaud E *et al.* 2011 *Phys. Lett.* **B702** 329
[5] Ahmed Z *et al.* 2011 *Phys. Rev.* **D84** 011102(R)
[6] Armengaud E *et al.* 2012 *Phys. Rev.* **D86** 051701(R)
[7] Savage C, Gelmini G, Gondolo P and Freese K 2009 *JCAP* **04** 010
[8] Angloher G *et al.* 2012 *Eur. Phys. J.* C 72 1971
[9] Aalseth C.E *et al.* 2013 *Phys. Rev.* D**88** 012002
[10] Armengaud E *et al.* 2013 *JCAP* **11** 67
[11] Gascon J and Bastidon N *et al.* 2014 *J Low Temp Phys* 176:870

Journal of Physics: Conference Series **606** (2015) 012003 doi:10.1088/1742-6596/606/1/012003

Dark matter direct detection with SuperCDMS Soudan

D. Speller for the SuperCDMS Collaboration
333 LeConte Hall, Department of Physics, University of California at Berkeley, Berkeley, CA, 94720 USA

E-mail: speller@berkeley.edu

Abstract. The Cryogenic Dark Matter Search Experiment (CDMS) and its successor, SuperCDMS, have had a long history of establishing world-leading upper limits on the interaction of Weakly Interacting Massive Particles (WIMPs) with standard model nucleons. SuperCDMS uses arrays of cryogenic germanium detectors to achieve excellent discrimination between the nuclear recoils expected for WIMP interactions and radioactively produced electron recoils through the collection of ionization and athermal phonons. Recent analyses of the data collected from the current installation of Interleaved Z-Sensitive Ionization and Phonon (iZIP) detectors have probed new regions of the WIMP parameter space, particularly in the region of low-mass WIMP interactions. This paper will discuss the current work of the SuperCDMS collaboration.

1. Motivation

There is strong evidence that approximately 27% of the mass-energy density of the universe is comprised of an invisible, non-interactive component commonly known as "dark" matter [1]. Astrophysical and cosmological constraints require that the dark matter be cold, neutral, stable, and massive, with weak scale interactions [2, 3], while compelling evidence from cluster merger 1E0657-558 (the "Bullet Cluster") provides support for the argument that the majority of dark matter is also non-baryonic [4]. Extensions to the standard model of particle physics provide several promising candidates, including a class of supersymmetric candidates collectively known as "Weakly Interacting Massive Particles", or WIMPs [3, 5].

2. CDMS, SuperCDMS and the iZIP detector

The Cryogenic Dark Matter Search Experiment (CDMS) [6, 7, 8, 9, 10] seeks to directly detect nuclear recoils from the elastic scattering of WIMP dark matter on the nuclei of semiconducting crystals cryogenically cooled to ~50 mK. CDMS uses Z-sensitive Ionization and Phonon (ZIP) detectors (Figure 1, left) to perform simultaneous measurements of ionization and athermal phonon signals, for event-by-event discrimination in the removal of backgrounds. For each event, the ionization yield (Figure 1, right) can be defined as the ratio of ionization to phonons,

$$Y = \frac{Q}{P}, \tag{1}$$

where Q is the collected ionization energy and P is the phonon recoil energy. The suppression of ionization yield in nuclear interactions relative to electron interactions provides the ability to

Journal of Physics: Conference Series **606** (2015) 012003 doi:10.1088/1742-6596/606/1/012003

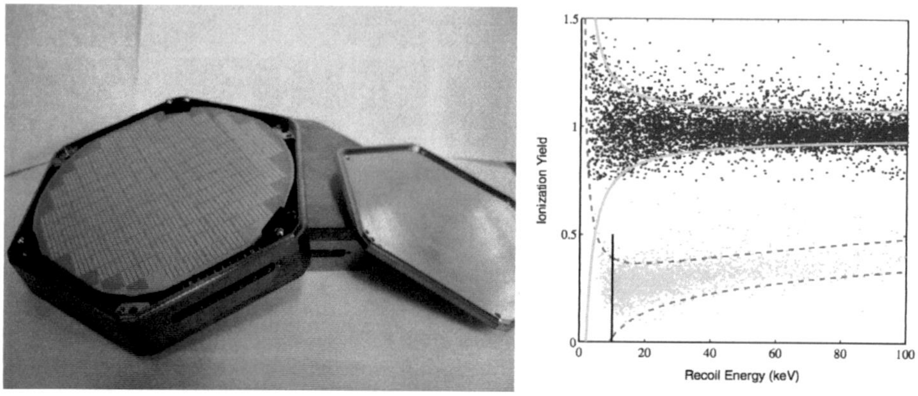

Figure 1. Left: CDMS II style ZIP detector in housing. Right: Ionization yield versus recoil energy for calibration data with a Cf-252 gamma and neutron source for select ZIP detectors in CDMS II Tower 1. During analysis, the ionization yield for electron recoils is nomalized to $Y = 1$. Figure from [6].

discern between nuclear and electron recoils, in turn establishing a significant handle for removing backgrounds from the analysis. The expected scenario in the CDMS detection paradigm is the following. An entering WIMP scatters from an electron or nucleus within the detector, resulting in the release of ionization and phonons within the crystal. Application of a bias voltage across the crystal allows collection of the released charge carriers. Meanwhile, athermal (recoil and Luke-Neganov) phonons are collected by arrays of superconducting tungsten Transition Edge Sensors (TES) [11]. Detailed descriptions of the readout chain and data acquisition systems have been addressed in past publications (see, for example, [12]).

While ionization yield is an excellent discriminator, incomplete ionization collection can lead to the leakage of background events into the nuclear recoil region. In particular, past challenges include leakage from surface events on the top and bottom of the detector, where the charge collection faces attrition due to electronic carrier trapping [13]; and sidewall events, where trapping may occur on the outer walls of the detector and the charge is not fully collected on the ionization electrode. This challenge was especially prominent in the detectors of CDMS II, which collected ionization on one side and phonons on the opposite side (Figure 1, left)[6].

Beginning with the installation of SuperCDMS, the detector design was significantly modified to address the surface event backgrounds faced by CDMS and CDMS II. The Interleaved Z-sensitive Ionization and Phonon (iZIP) detector was developed in part to address the issue of incomplete ionization collection by allowing collection of both charge and phonons on both sides of the detector (Figure 2). Each face of the detector contains 4 phonon readout channels: 3 inner channels and an outer guard ring. Each side also contains an inner and outer electrode for charge fiducialization. Ionization and phonons are collected on both sides of the detector; biasing of the charge sensors determines the charge carrier collected on each side. Surface events are collected completely on one side of the detector. A uniform field is established in the bulk of the crystal, drifting charge carriers symmetrically to both sides (Figure 3). Surface event discrimination and identification of events collected in the outer electrode allow the definition of an optimal fiducial volume for analysis and enable an efficient separation of surface and bulk events. The interleaving of the phonon and ionization rails creates a complex electric field concentrated at the surfaces of the detector, such that events occurring near a surface are entirely collected on one side of the detector.

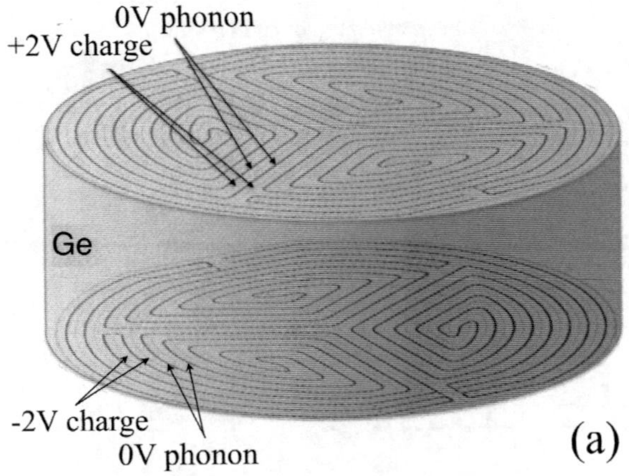

Figure 2. Top-side layout of the iZIP detector. Note the "mercedes" geometry of the interior channels and the interleaving of the ionization and phonon rails. The ionization lines and phonon sensors on each side are interleaved with a 1 mm pitch. The detectors are operated with a 4 volt bias across the crystal, with ionization lines on side 1 biased at +2V and side 2 biased at -2V, while the phonons sensors are 0 volts. The detectors are 76 mm in diameter and 25 mm thick. Figure from [14].

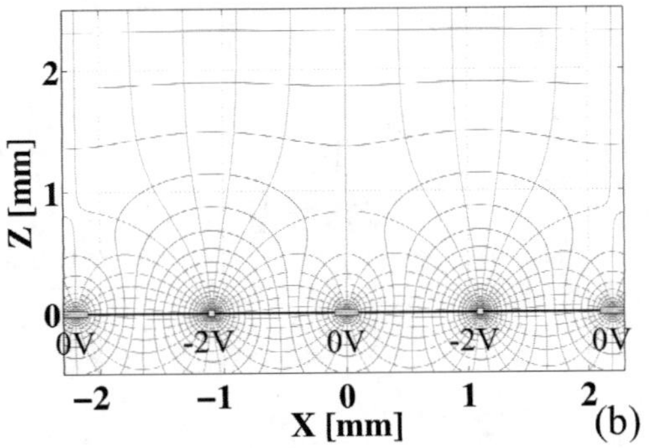

Figure 3. Electric field lines produced by interleaved ionization and phonon lines in the iZIP detector. Figure from [14].

3. SuperCDMS Soudan Installation

SuperCDMS is an upgrade to the CDMS and CDMS II experiments, and is currently running as a five-tower installation of germanium iZIP detectors in the Soudan Underground Laboratory (SUL) in Soudan, MN, located in an iron mine 2341 feet underground. SuperCDMS employs both passive and active shielding (Figure 4) to reduce external photon and neutron backgrounds from radioactive and cosmogenic sources, and is surrounded by successive layers of copper ($\sim$0.5 cm), lead (22.5 cm), and polyethylene (50 cm) [6]. The shielding is surrounded by a 5-cm-thick muon scintillator veto, which identifies any influx of charged particles (in addition to some neutral particles). Furthermore, the mine in which SUL is housed provides a 780 meter rock overburden, which is 2090 meters water equivalent (mwe). The latter reduces the surface muon flux by a factor of 5×10^4 [6]. Nuclear recoils due to external neutrons are well suppressed by the

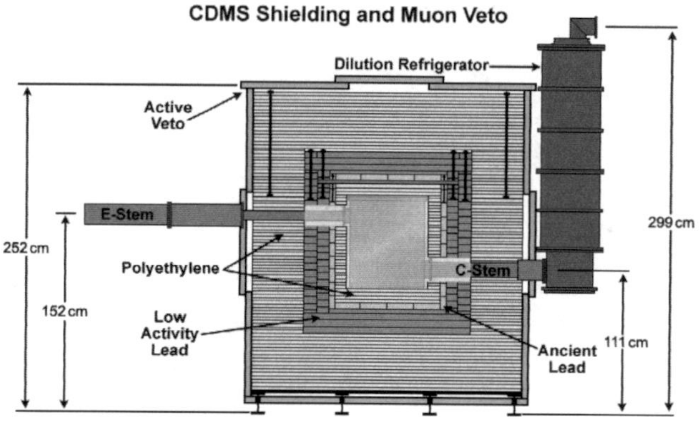

Figure 4. Side view of the shielding and muon veto from the SuperCDMS experiment. From [22].

Figure 5. iZIP detector. Figure from [14].

combination of active and passive shielding and the efficient tagging of incoming muons. Further details on the icebox and cryogenics of the SuperCDMS installation at SUL are discussed in, for example, [22].

4. Current work, recent studies, and analysis results

In continuation of the search for WIMPs, there have been a number of ongoing and recently published studies by the SuperCDMS collaboration involving the CDMS II germanium ZIP detectors and the currently installed iZIP detectors of SuperCDMS Soudan.

4.1. Continuing studies

CDMS II reanalysis SuperCDMS has undertaken a reanalysis of the data from CDMS II. This reanalysis was prompted by the discovery of an issue in the algorithm for the reconstruction of ionization pulses which required an upgrade in the software [23]. The publication for this

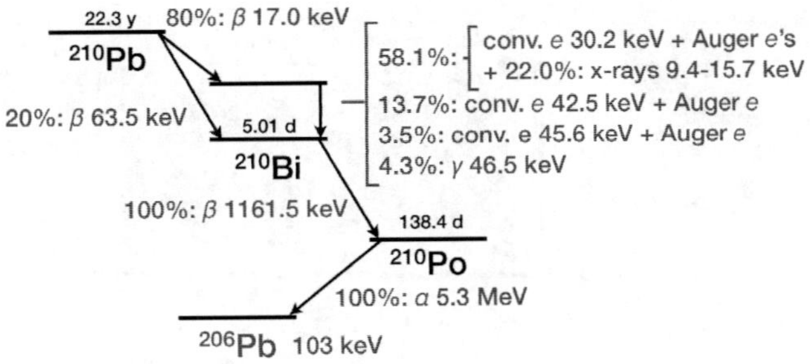

Figure 6. Decay chain for Pb-210. Figure from [14].

analysis is currently in preparation.

Search for annual modulation SuperCDMS is searching for an annual modulation in the data from CDMS II. Analysis of the data above a threshold of 5 keV_{nr} (i.e. keV, nuclear recoil equivalent energy; threshold set by the trigger efficiency) [16] yielded no significant modulation in the CDMS II low-energy data. Studies are underway to study the systematic uncertainties of an analysis extending to lower energies. This publication is also in preparation.

Nuclear recoil energy scale The nuclear recoil energy scale in Ge is important for understanding the energy scale of nuclear recoils in germanium. This analysis compares Monte Carlo to neutron calibration data. This publication is in preparation.

4.2. Low energy germanium likelihood analysis
In 2014, the SuperCDMS collaboration released a maximum-likelihood analysis of the low-energy germanium data in CDMS II [17] using a combination of GEANT4 Monte Carlo and calibration data to model the detector backgrounds. In contrast to an earlier, external analysis of the low-energy germanium data, which claimed a significant excess in the nuclear recoil event rate, CDMS found no statistically significant WIMP component.

4.3. Demonstration of rejection for iZIP detectors
The surface rejection of the iZIP detectors (Figure 5) operated at SUL was demonstrated in a study of the effect of surface and housing contamination from Pb-210, a product of the radon decay chain [14] (Figure 6). Two Pb-210 sources were installed on side one and two of the top and bottom detectors, respectively, of iZIP Tower 3, yielding approximately 70 events per hour per source. Using detector T3Z1 with the lead source facing side 1, the rejection power of the iZIP was studied on approximately 2500 live-hours of Ba-133 calibration data (Figure 7), and established an upper limit on the leakage fraction of less than 1.26×10^{-5} at the 90% confidence level between 8-115 keV_{nr}. One low-yield outlier was observed in the ionization yield plane which was easily removed with the application of a phonon symmetry cut. This study demonstrated the rejection requirement for the use of iZIP detectors at SNOLAB. See the article on "SuperCDMS SNOLAB Dark Matter Search" by P. Brink, this proceedings.

4.4. SuperCDMS Soudan low threshold analysis
The SuperCDMS low threshold analysis [18] was conducted using 7 low-threshold detectors selected from the current 5-tower iZIP installation in SUL. A blind analysis was performed on

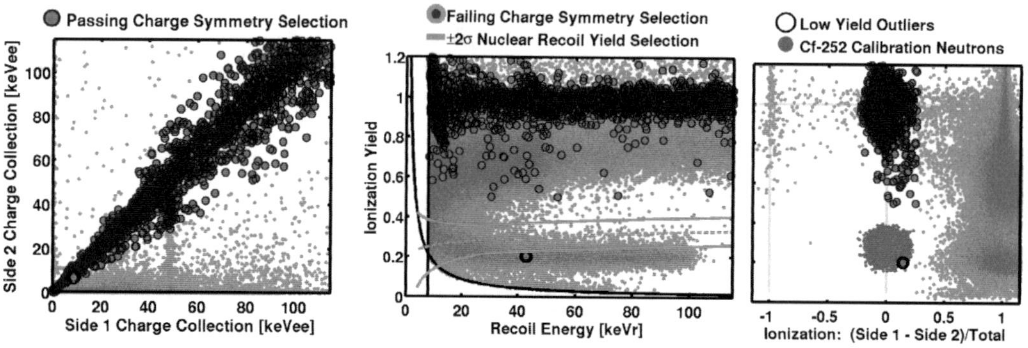

Figure 7. Left: Charge symmetry selection of bulk events in iZIP detector. Blue points pass the symmetry cut (approximately equal energy deposition on both sides of the detector); red points fail the charge symmetry selection. Middle: Ionization yield versus recoil energy of Ba-133 calibration data. The red points (surface events) failed the charge symmetry selection shown on the left. Blue points pass the charge symmetry selection. One low yield outlier is easily removed via application of a phonon symmetry cut. The green band shows the $\pm 2 - \sigma$ nuclear recoil signal region, based on calibration data from Cf-252. Right: Charge partition of Ba-133 events. Electron recoils and surface events are well separated from the signal manifold. Figures from [24].

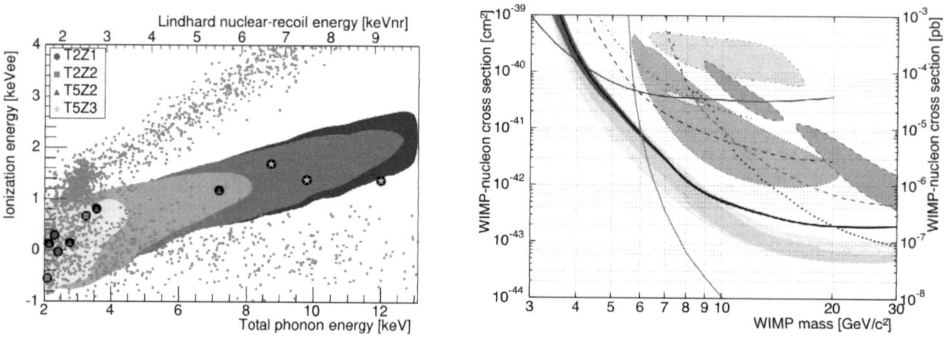

Figure 8. Left: Candidate events from the SuperCDMS low threshold analysis (2014) [18]. Right: 90 % Upper-limit on SI WIMP interactions, 95 % C.L (black). Figure from [18].

a 577 kg-day exposure, using a boosted decision tree (BDT) approach to search for light-mass WIMPs. At the time of publication (Figure 8), this analysis was the most senstive search for light-mass WIMPs between approximately 3.5-6 keV_{nr}, establishing a new upper-limit of 1.2×10^{-42} cm^2 at 8 GeV/c^2.

4.5. CDMSlite

In CDMSlite, a single iZIP detector ($\sim$0.6 kg) is operated in a special one-sided configuration at high voltage, allowing for the Luke-Neganov amplification of ionization energy deposited during detector interactions [19]. This configuration attained a low energy threshold of 170 keV_{ee} and allows previously unprecedented sensitivity to low-mass WIMP interactions, establishing a new upper-limit of $\sim 4.5 \times 10^{-42}$ cm^2 at 6 GeV/c^2 (Figure 9).

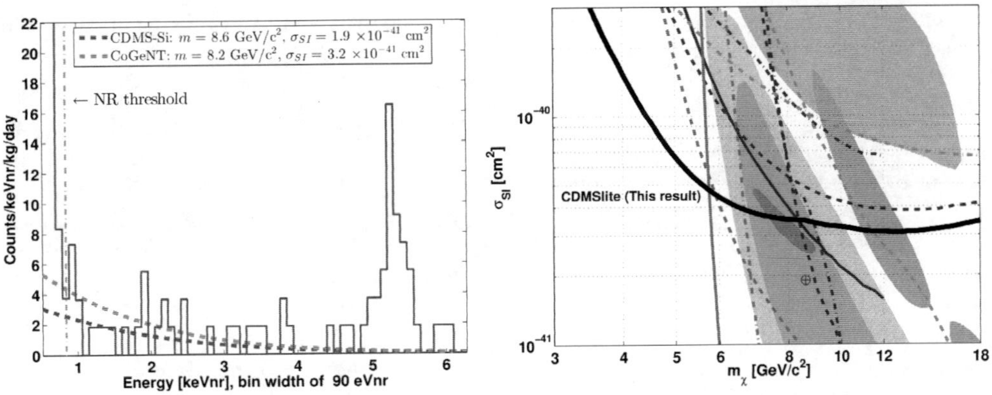

Figure 9. Left: CDMSlite spectrum, with comparison to CDMS-Si and CoGeNT. Right: 90% Upper confidence limit on SI-independent WIMP interactions set by CDMSlite (black). Figures from [19].

4.6. SuperCDMS high mass WIMP search

Currently, SuperCDMS is engaged in the analysis of the first high-threshold results from the 5-tower iZIP installation at SUL. SuperCDMS has operated a complement of 15 detectors for ∼2 years of low-background running, interspersed with periods of Ba-133 and Cf-252 calibration data. The analysis is currently underway. Related studies include the work of J. Morales, "Understanding sources of noise in the Ge detectors of the CDMS experiment", in this proceedings. Furthermore, significant progress is being made in the modeling of backgrounds and detector response using GEANT4 and the detector Monte Carlo. Additional details on this work will be discussed in future publications. Finally, with the allowed regions of minimal extensions to supersymmetry (SUSY) becoming increasingly constrained by experimental limits, a number of published works have begun to address the experimental implications of additional WIMP operators beyond the standard SD/SI interactions (see, for example, [20]). Several members of the collaboration have begun to explore a broader range of WIMP nucleus interactions describable by effective field theory (EFT) in the context of SuperCDMS [26].

4.7. Charge Transport

As detector sensitivity requirements become more demanding, a thorough understanding of detector behavior, including the characteristics of charge transport in germanium, becomes increasingly important. Studies on charge transport in high purity, low temperature germanium are discussed in the work of A. Phipps ("Charge carrier transport and capture in low temperature, high purity germanium") and R. Moffatt ("Two-dimensional spatial imaging of charge transport in germanium at low temperature"), this proceedings.

4.8. R&D: Towards SuperCDMS SNOLAB

Research and development towards the next generation of cryogenic dark matter detectors has been conducted in parallel with current SuperCDMS operations. In the summer of 2014, SuperCDMS SNOLAB was selected as one of three primary generation-2 dark matter search experiments. A full discussion can be found in in this proceedings in the work of P. Brink ("SuperCDMS SNOLAB Dark Matter Search"). In addition, performance measurements of the next generation 100-mm iZIP detectors conducted at the SuperCDMS test facility at the University of Minnesota is presented in the work of A. Kennedy ("Initial phonon and ionization measurements of 100-mm-diameter germanium dark matter detectors for use in SuperCDMS SNOLAB"), also in this proceedings.

5. Conclusion

The SuperCDMS collaboration is currently engaged in a full program of of analysis, operations, and research and development for the next generation of dark matter direct detection through the use of high purity, low temperature germanium detectors. Analysis of the approximately two years of data collected with the currently installed iZIP payload is underway, while the detectors remain cold and running at Soudan Underground Lab. Continuing analyses of data collected with CDMS detectors have made significant progress, with several publications in preparation, along with concerted efforts to enhance understanding of the backgrounds and detector characteristics. We are optimistic that the cryogenic germanium detectors of the SuperCDMS experiment will continue to play a major role in the future of dark matter direct detection.

Acknowledgments

The SuperCDMS collaboration gratefully acknowledges the contributions of numerous engineers and technicians, in addition to the assistance from the staff of the Soudan Underground Laboratory and the Minnesota Department of Natural Resources. Part of the research for CDMSlite was conducted under the Ultra Sensitive Nuclear Measurements Initiative at Pacific Northwest National Laboratory, which is operated by Battelle for the U.S. Department of Energy. The iZIP detectors were fabricated in the Stanford Nanofabrication Facility, which is a member of the National Nanofabrication Infrastructure Network. This work is supported in part by the National Science Foundation, by the United States Department of Energy, by NSERC Canada, and by MultiDark (Spanish MINECO). Fermilab is operated by the Fermi Research Alliance, LLC under Contract No. De-AC02-07CH11359. SLAC is operated under Contract No. DE-AC02-76SF00515 with the United States Department of Energy.

References

[1] P A R Ade e a 2013 *Astronomy & Astrophysics* **571** A16
[2] Trimble V 1987 *Annu. Rev. Astro. Astrophys.* **25** 425–72
[3] Bergstrom L 2000 *Rep. Prog. Phys.* **63** 793 (*Preprint* `http://arxiv.org/pdf/hep-ph/0002126v1.pdf`)
[4] Clowe D, Bradac M, Gonzalez A, Markevitch M, Randall S, Jones C and Zaritsky D 2006 *The Astrophysical Journal* **648** L109–L113
[5] Primack J R, Seckel D and Sadoulet B 1988 *Annu. Rev. Nucl. Part. Sci.* **38** 751–807
[6] Akerib D S, Alvaro-Dean J, Armel-Funkhouser M, Attisha M, Baudis L, Bauer D, Beaty J, Brink P, Bunker R, Burke S P and et al B C 2004 *Phys. Rev. Lett* **93** 211301
[7] Collaboration C 2000 *Phys. Rev. Lett.* **84**(25) 5699–5703 URL `http://link.aps.org/doi/10.1103/PhysRevLett.84.5699`
[8] Collaboration C 2002 *Phys. Rev. D* **66**(12) 122003 URL `http://link.aps.org/doi/10.1103/PhysRevD.66.122003`
[9] Collaboration C 2003 *Phys. Rev. D* **68** 082002 URL `http://journals.aps.org/prd/pdf/10.1103/PhysRevD.68.082002`
[10] Collaboration C I 2010 *Science* **327** 1619–1621
[11] Cabrera B, Clarke R, Miller A, Nam S W, Romani R, Saab T and Young B 1999 *To appear in Proceedings of the 22nd International Conference on Low-Temperature Physics*
[12] Bauer D and et al 2011 *Nuclear Instruments and Methods in Physics Research A* **638**
[13] Pyle M 2012 *Optimizing the Design and Analysis of Cryogenic Semiconductor Dark Matter Detectors for Maximum Sensitivity* Ph.D. thesis Stanford University
[14] Collaboration C 2013 (*Preprint* `ins-det/1305.2405`)
[15] Hertel S 2010 Internal document
[16] Collaboration C 2012 (*Preprint* `astro-ph/1203.1309`)
[17] Collaboration S 2014 (*Preprint* `astro-ph/1410.1003`)
[18] Collaboration S 2014 *Phys. Rev. Lett.* **112** 241302
[19] Collaboration S 2014 *Phys. Rev. Lett.* **112** 041302
[20] Fitzpatrick A L and et al 2013 *JCAP* **1302**

[21] Jungman G, Kamionkowski M and Griest K 1996 *Physics Reports* **267** 238 (*Preprint* hep-ph/9506380)
[22] Sander J 2007 *Results from the Cryogenic Dark Matter Search Using a Chi Squared Analysis* Ph.D. thesis University of California at Santa Barbara
[23] Kiveni J 2009 *A Search for WIMP Dark Matter Using an Optimized Chi-square Technique on the Final Data from the Cryogenic Dark Matter Search Experiment (CDMS II)* Ph.D. thesis Syracuse University
[24] Collaboration S 2013 Proposal
[25] Pyle M, Brink P, Cabrera B, Castle J, Colling P, Chang C, Cooley J, Lipus T, Ogburn R W and Young B 2006 *Nuclear Instruments and Methods in Physics Research A* **559** 405–407
[26] Schneck K, Cabrera B, Cerdeno D G, Mandic V and Rodgers H e a 2015 (*Preprint* arXiv:1503.03379)

The MAJORANA DEMONSTRATOR: A Search for Neutrinoless Double-beta Decay of ^{76}Ge

W. Xu[1], N. Abgrall[2], F. T. Avignone III[3,4], A. S. Barabash[5], F. E. Bertrand[4], V. Brudanin[6], M. Busch[7,8], M. Buuck[9], D. Byram[10], A.S. Caldwell[11], Y-D. Chan[2], C. D. Christofferson[11], C. Cuesta[9], J. A. Detwiler[9], Yu. Efremenko[12], H. Ejiri[13], S. R. Elliott[1], A. Galindo-Uribarri[4], G. K. Giovanetti[14,8], J. Goett[1], M. P. Green[4], J. Gruszko[9], I. Guinn[9], V. E. Guiseppe[3], R. Henning[14,8], E. W. Hoppe[15], S. Howard[11], M. A. Howe[14,8], B. R. Jasinski[10], K. J. Keeter[16], M. F. Kidd[17], S. I. Konovalov[5], R. T. Kouzes[15], B. D. LaFerriere[15], J. Leon[9], J. MacMullin[14,8], R. D. Martin[10], S. J. Meijer[14,8], S. Mertens[2], J. L. Orrell[15], C. O'Shaughnessy[14,8], N. R. Overman[15], A. W. P. Poon[2], D. C. Radford[4], J. Rager[14,8], K. Rielage[1], R. G. H. Robertson[9], E. Romero-Romero[12,4], M. C. Ronquest[1], B. Shanks[14,8], M. Shirchenko[6], N. Snyder[10], A. M. Suriano[11], D. Tedeschi[3], J. E. Trimble[14,8], R. L. Varner[4], S. Vasilyev[6], K. Vetter[2 18], K. Vorren[14,8], B. R. White[4], J. F. Wilkerson[14,8,4], C. Wiseman[3], E. Yakushev[6], C-H. Yu[4], and V. Yumatov[5]

The MAJORANA Collaboration

[1]Los Alamos National Laboratory, Los Alamos, NM, USA
[2]Nuclear Science Division, Lawrence Berkeley National Laboratory, Berkeley, CA, USA
[3]Department of Physics and Astronomy, University of South Carolina, Columbia, SC, USA
[4]Oak Ridge National Laboratory, Oak Ridge, TN, USA
[5]Institute for Theoretical and Experimental Physics, Moscow, Russia
[6]Joint Institute for Nuclear Research, Dubna, Russia
[7]Department of Physics, Duke University, Durham, NC, USA
[8]Triangle Universities Nuclear Laboratory, Durham, NC, USA
[9]Center for Experimental Nuclear Physics and Astrophysics and Department of Physics, University of Washington, Seattle, WA, USA
[10]Department of Physics, University of South Dakota, Vermillion, SD, USA
[11]South Dakota School of Mines and Technology, Rapid City, SD, USA
[12]Department of Physics and Astronomy, University of Tennessee, Knoxville, TN, USA
[13]Research Center for Nuclear Physics and Department of Physics, Osaka University, Ibaraki, Osaka, Japan
[14]Department of Physics and Astronomy, University of North Carolina, Chapel Hill, NC, USA
[15]Pacific Northwest National Laboratory, Richland, WA, USA
[16]Department of Physics, Black Hills State University, Spearfish, SD, USA
[17]Tennessee Tech University, Cookeville, TN, USA

E-mail: wxu@lanl.gov

[18] Alternate Address: Department of Nuclear Engineering, University of California, Berkeley, CA, USA

Journal of Physics: Conference Series **606** (2015) 012004 doi:10.1088/1742-6596/606/1/012004

Abstract. Neutrinoless double-beta ($0\nu\beta\beta$) decay is a hypothesized process where in some even-even nuclei it might be possible for two neutrons to simultaneously decay into two protons and two electrons without emitting neutrinos. This is possible only if neutrinos are Majorana particles, *i.e.* fermions that are their own antiparticles. Neutrinos being Majorana particles would explicitly violate lepton number conservation, and might play a role in the matter-antimatter asymmetry in the universe. The observation of neutrinoless double-beta decay would also provide complementary information related to neutrino masses. The MAJORANA Collaboration is constructing the MAJORANA DEMONSTRATOR, with a total of 40-kg Germanium detectors, to search for the $0\nu\beta\beta$ decay of ^{76}Ge and to demonstrate a background rate at or below 3 counts/(ROI·t·y) in the 4 keV region of interest (ROI) around the 2039 keV Q-value for ^{76}Ge $0\nu\beta\beta$ decay. In this paper, we discuss the physics of neutrinoless double beta decay and then focus on the MAJORANA DEMONSTRATOR, including its design and approach to achieve ultra-low backgrounds and the status of the experiment.

1. Introduction

Since its discovery and confirmation by the Super-Kamiokande [1], the Sudbury Neutrino Observatory (SNO) [2] and other experiments [3], a non-zero neutrino mass has been recognized as an indication of physics beyond the Standard Model of particle physics, and therefore one of the most fundamental questions. Majorana particles are fermions that are their own anti-particles. If neutrinos are Majorana particles, neutrinoless double-beta ($0\nu\beta\beta$) decay can take place in some even-even nuclei where single-beta decay is energetically forbidden or highly suppressed. In this process, two neutrons simultaneously decay into two protons and two electrons without emitting neutrinos. This would explicitly violate the lepton number conservation [4, 5, 6]. $0\nu\beta\beta$ decay experiments are the only way to unambiguously establish the Majorana or Dirac nature of standard model neutrinos and the observation of $0\nu\beta\beta$ decay would have a profound impact.

$0\nu\beta\beta$ decay requires neutrinos to be massive and their own antiparticles. A direct relationship can be established between $0\nu\beta\beta$ decay half time and the effective Majorana neutrino mass as

$$(T^{0\nu}_{1/2})^{-1} = G^{0\nu}|M_{0\nu}|^2\left(\frac{< m_{\beta\beta} >}{m_e}\right)^2, \tag{1}$$

where $G^{0\nu}$ is the phase space of the decay, $M_{0\nu}$ is the nuclear matrix element and $< m_{\beta\beta} >= |\Sigma^3_{i=1}U^2_{ei}m_i|$ is the effective Majorana neutrino mass with U_{ei} being the Pontecorvo-Maki-Nakagawa-Sakata mixing matrix [7]. Therefore, measurement of the $0\nu\beta\beta$ decay half time is a complementary way to probe the absolute neutrino mass scale. For example, the inverted mass hierarchy has an effective neutrino mass between about 15 meV and 50 meV, and the lower boundary corresponds to a ^{76}Ge $0\nu\beta\beta$ decay half time longer than 10^{27} years.

In recent decades, many experiments have searched for $0\nu\beta\beta$ decay in several isotopes including, *e.g.*, ^{76}Ge, ^{136}Xe, ^{130}Te, and others. For recent reviews of $0\nu\beta\beta$ decay experiments, see for example references [6, 8]. All of these isotopes also undergo two-neutrino double-beta ($2\nu\beta\beta$) decay with typical half-lives on the order of 10^{19} to 10^{21} years [9, 10, 11, 8]. Similar to a single-beta decay spectrum, the sum energy of the two electrons in $2\nu\beta\beta$ decay is a continuum extending up to the Q-value ($Q_{\beta\beta}$) of the decay. In contrast, due to the absence of neutrinos, the two electrons in $0\nu\beta\beta$ decay carry almost all of the energy released in the decay, and their total kinetic energy is peaked at the Q-value. Compared to other detection systems, High-Purity Germanium (HPGe) detectors have the critical advantage of excellent energy resolution, which eliminates the otherwise almost irreducible $2\nu\beta\beta$ decay background and improves the discovery potential due to the clean line signature.

2. The Majorana Demonstrator

The Majorana Collaboration is constructing the Majorana Demonstrator (MJD) [7] in class-100 clean rooms located on the 4850' underground level at the Sanford Underground Research Facility (SURF) [12]. The collaboration plans to deploy a total of 40-kg Germanium detectors into two modules, with 30 kg of the Ge detectors enriched to 87% ^{76}Ge. The Demonstrator aims at demonstrating a path forward to achieve a ultra low background and at showing technical and engineering scalability toward a tonne-scale instrument.

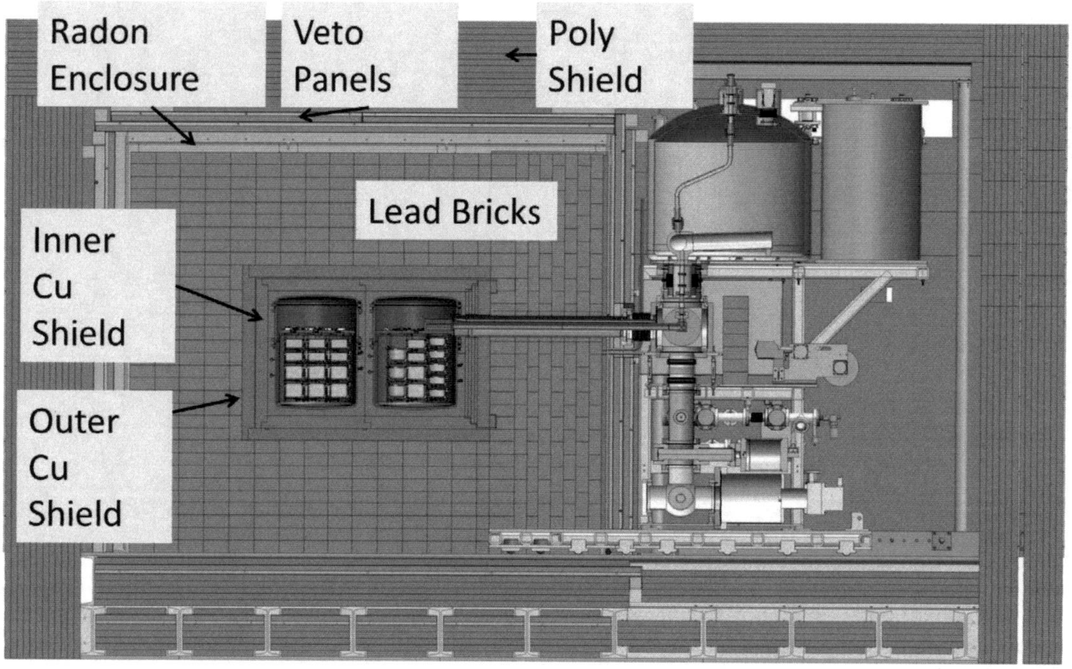

Figure 1. Adapted from [7]. MJD shield configuration. The experimental apparatus is sitting on the top of a stainless-steel over-floor with orthogonal channels, allowing HDPE panels (poly shield) and veto panels underneath the apparatus. See the text for detailed discussions of the shield.

To be specific, a background rate at or below 1.0 count/(ROI·t·y) in the 4 keV region of interest (ROI) around the 2039 keV Q-value for ^{76}Ge $0\nu\beta\beta$ decay is required for tonne-scale Ge-based experiments that will probe the inverted hierarchy parameter space for $0\nu\beta\beta$ decay. Based on simulation studies, if the Demonstrator can achieve a background level of 3.0 counts/(ROI·t·y) or better, then a tonne-scale experiment with similar design and material is expected to have the required background level of 1.0 count/(ROI·t·y) or better, thanks to various effects such as self-shielding and longer time for cosmogenic backgrounds such as ^{68}Ge to decay. In addition to searching for $0\nu\beta\beta$ decay, the Demonstrator will be used to search for a range of new physics beyond the standard model, including low mass dark matter and axions [7, 13].

A modular approach is chosen for MJD for its natural expandability to larger scale experiment. In the current approach, four or five HPGe detectors are stacked together to form a string, and seven strings are mounted into a single cryostat with dedicated supporting systems such as cryogenic and vacuum systems, together referred to as a module. The construction of the DEMONSTRATOR is organized in three phases. First, a prototype module with three strings of natural HPGe detectors was constructed in 2014 and has been in commissioning. By early 2015, the first production module with more than half of the total enriched HPGe detectors and some natural detectors, a total mass of 20 kg of Ge, will be completed. By late 2015, the second production module with the rest of HPGe detectors will be constructed, also with a total mass of 20 kg of Ge. Thanks to separate supporting systems, those modules can be individually operated and maintained, and data-taking will begin as soon as the first module is ready. The rest of this proceedings will report on the status of the MJD project as of September 2014.

Figure 2. Left: A photograph of the main body of the lead shield and outer copper shield. Right: A photograph of the nearly completed shield, inside which the prototype module has been taking commissioning data. The supporting systems of the prototype module can be seen in the photo as well. The cryostat of the prototype module is inside the shield. The current prototype module shielding includes large veto coverage, a semi-sealed Rn box constantly purged with nitrogen gas, complete lead shield at the prototype module side, and outer copper shield surrounding the cryostat.

2.1. Shield

MJD utilizes both active and passive shielding, as shown in Fig. 1. The outmost component of the shield configuration is composed of layers of high density polyethylene (HDPE) panels, two layers of which are borated. The construction of the poly shield has started, and upon completion, it will enclose the entire apparatus, including the supporting systems of each module. Inside the poly shield, there are two layers of active veto panels on every side of the experiment, made of scintillating acrylic sheet and read out by Photomultiplier Tubes (PMTs). The active veto panels on four sides of the shield have been installed and commissioned. Inside the veto panels, a semi-sealed Radon (Rn) Enclosure Box made of Aluminum has been constructed to provide a nearly Rn-free environment. A nitrogen gas delivery system that constantly purges the Rn box with boil-off nitrogen gas was implemented. Further improvements are currently being implemented to use cold clean charcoal to filter out the remaining Rn in the nitrogen purge gas. Inside the Rn box is a 45 cm thick lead shield composed of $5.1 \times 10.2 \times 20.3$ cm^3 lead bricks, which are carefully stacked in a pattern that eliminates any direct path for photons originating from outside the shield to penetrate. The main body of the lead shield has already been completed,

Journal of Physics: Conference Series **606** (2015) 012004 doi:10.1088/1742-6596/606/1/012004

as shown in the left side of Fig. 2. In the photo, keyed structures can be seen on two sides of the lead shield. When cryostats are inserted into the shield, additional lead shielding matching the keyed structures will be also installed, completing the lead shield. Inside the lead shield, an outer copper shield made of Oxygen-Free High thermal Conductivity (OFHC) copper has already been installed. An inner copper shield made of electroformed copper will be installed to directly enclose the cryostats. For more details of the shield design, see reference [7]. Since August 2014, the MJD prototype module has been taking commissioning data in the nearly completed shield described here, as shown in the photo on the right side of Fig. 2.

2.2. Detectors, Strings, and Modules

The collaboration has chosen to use P-type Point Contact (P-PC) HPGe detectors [14, 15]. As compared to the more conventional coaxial HPGe detectors, they have several important advantages for a rare-event search. Their small readout contact produces a highly localized "weighting potential", and this combined with a larger range of charge drift times allows for excellent pulse-shape discrimination sensitivity to distinguish between multi-site and single-site events. This suppresses the main background of Compton-scattered gamma rays. Both P-PC and p-type coaxial detectors also have a thick outer lithium contact that fully absorbs alpha particles, further reducing the background. Lastly, the small capacitance of PPC detectors greatly improves the energy resolution at low energies, reducing the energy threshold and allowing for low-mass dark matter searches.

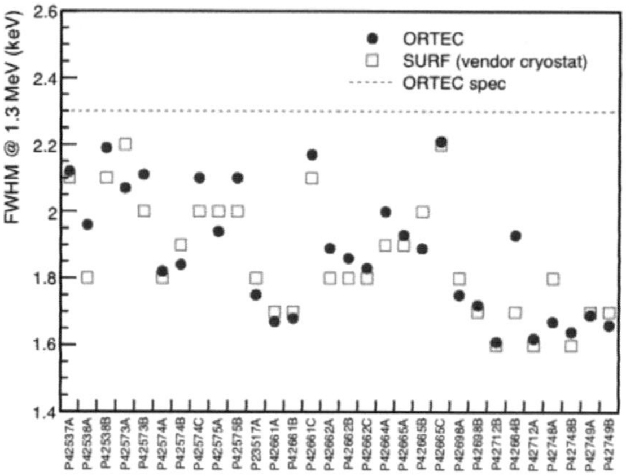

Figure 3. Left: A photograph of a digital microscope that has been used to measure the dimensions of bare Ge crystals inside a custom glove box. A digital scale is used for mass measurements. Right: The Full-Width-at-Half-Maximum (FWHM) at 1333 keV of the 30 enriched ORTEC detectors measured by ORTEC (blue dots) and by the MJD collaboration at SURF (red open squares), plotted against detector serial number. The FWHM of all detectors are better than the experimental specification of 2.3 keV, which is shown as the dotted horizontal line.

Most of the natural Ge detectors being deployed at MJD are the Broad Energy Ge detectors (BEGe) manufactured by Canberra [16], and all of the enriched Ge detectors and two natural Ge detectors are manufactured by AMETEK/ORTEC [17]. A comprehensive set of acceptance and characterization tests of the ORTEC detectors have been carried out at SURF [18], including measurements of the masses and dimensions of bare Ge crystals, as shown in the left side photo

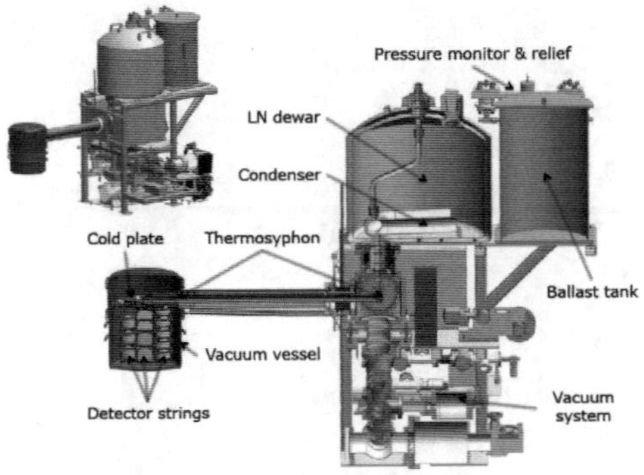

Figure 4. Left: Adapted from [7]. The design of a MAJORANA DEMONSTRATOR module. Right: A photograph of the prototype module during string mounting. The unseen cryostat is inside a custom glove box so that it can be opened up to allow string mounting. The visible copper tube that connects the cryostat and the vacuum system is called "cross-arm tube", within which is the thermosyphon tube. The thermosyphon tube is filled with nitrogen in dual phases. The copper structure surrounding the cross-arm tube is used to support part of the lead shielding, which completes the main body of the lead shield on the over-floor when the cryostat is inserted into the shield.

of Fig. 3. So far, 30 enriched detectors with a total mass of 25.2 kg have been delivered to SURF and all of them meet experimental specifications such as energy resolution, as shown in the right panel of Fig. 3. Natural BEGe detectors have also been tested. Efforts to manufacture more detectors to reach 30 kg are still ongoing.

Bare Ge crystals are removed from their commercial cryostats and then built into detector units packed with custom high voltage contacts and front-end signal read-out boards. Depending on the size of the bare crystals, four or five detector units are stacked into one string, and each string is then tested for performance. The metal supporting parts in the detector units and strings are made of ultra pure electroformed copper [19] produced underground to minimize radioactivity at the vicinity of detectors. More details about detector units, strings and their performance can be found in other MJD talks in this workshop: I. Guinn, Low Background Signal Electronics for the MAJORANA DEMONSTRATOR; J. Leon, Prototype for a low-noise charge preamplifier with a forward-biased JFET; S. Mertens, MAJORANA Experience with Germanium Detectors; and B. Jasinski, Assembly of MJD.

A total of fourteen production strings of HPGe detectors will be constructed, seven for each production module. The construction of strings for the first module is well underway as of December 2014. In addition, three strings with a total of ten natural HPGe detectors of both BEGe and ORTEC types have been constructed for the prototype module and are being used for commissioning of the experimental apparatus.

Each MJD module consists of a cryostat with detector strings, cryogenic and vacuum systems, parts of the shield and its own calibration system, as shown in the design plot in the left panel of

Journal of Physics: Conference Series **606** (2015) 012004 doi:10.1088/1742-6596/606/1/012004

Fig. 4. Each module can be moved as a whole inside the clean room. The cryogenic system is a thermosyphon filled with nitrogen in both gas and liquid phases. The thermosyphon thermally connects to a cold plate inside the cryostat at one end and it is constantly cooled inside a Liquid Nitrogen (LN) dewar at the other end [20]. For production modules, the cryostat itself and the thermosyphon are all made of electroformed copper.

To validate the apparatus design and to debug issues in the construction and operation of modules and the DAQ systems, a prototype module was constructed and has been in commissioning, as shown in the right panel of Fig. 4. The prototype module is identical to production modules, except it is made of commercial OFHC copper. The vacuum system of the prototype module will be reused for the second production module. Ten natural HPGe detectors of both BEGe and ORTEC types are arranged into three strings and mounted onto the cold plate inside the prototype module cryostat, as shown in the left panel of Fig. 5.

DAQ systems based on the Object-Oriented Real-time Control and Acquisition (ORCA) platform [21] have been instrumented to control both commercial and collaboration-manufactured electronics, and background data and calibration data are being taken. The HPGe detectors in the prototype cryostat have shown good energy resolution similar to that in commercial cryostats. The experience of commissioning the prototype module has been critical for the collaboration to improve the experimental apparatus. In addition, a set of detailed construction procedures and strict cleanness protocols, including clean machining, have been developed and improved during this process. The prototype module commissioning will continue for a few months and the construction of the first production module is well underway and will be completed in early 2015.

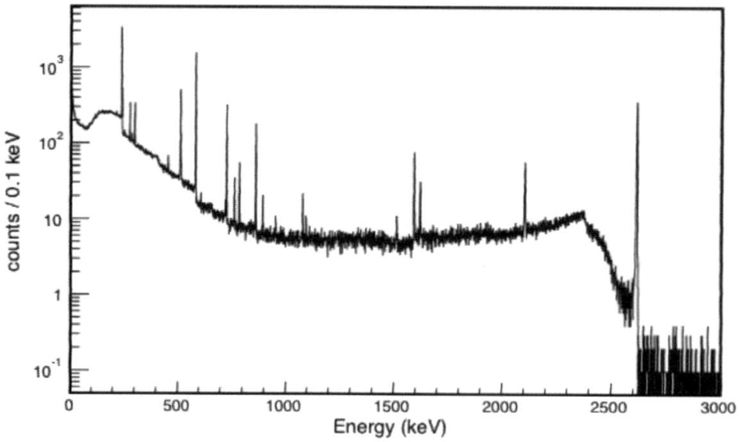

Figure 5. Left: A photograph inside the prototype cryostat, which is opened up to allow the mounting of three strings of HPGe detectors. The string in the front has two Canberra BEGe detectors on the top and two ORTEC detectors at the bottom. Right: A spectrum taken by one of the HPGe detectors in the prototype cryostat during in a calibration run with a ^{228}Th line source.

2.3. Background Estimation

A vigorous program to assay the natural radioactivity in almost all materials used in the MJD experiment has provided a foundation for accurate background projections, summarized in Fig. 6. The radioactivity of components close to the active HPGe detector volumes has a strong influence on the background level. In particular, ultra pure copper electroformed underground has both much reduced primordial radioactivity and limited cosmogenically produced ^{60}Co. By using electroformed copper at some locations closest to the detectors, such as the support structures in detector units and strings as well as the cryostat itself, we have achieved several orders-of-magnitude background reduction over the use of commercial alternatives. The biggest projected background is from natural uranium and thorium in the electronics front-end, cables and connectors, which are also very close to the detectors. Some of the estimations are upper limits and work to improve them are still in progress. Summarizing all the contributions, the projected background for MJD is $\leq$ 3.1 counts/(ROI-t-y) in the 4 keV ROI around 2039 keV. More details about the MJD background model can be found in references [7, 22].

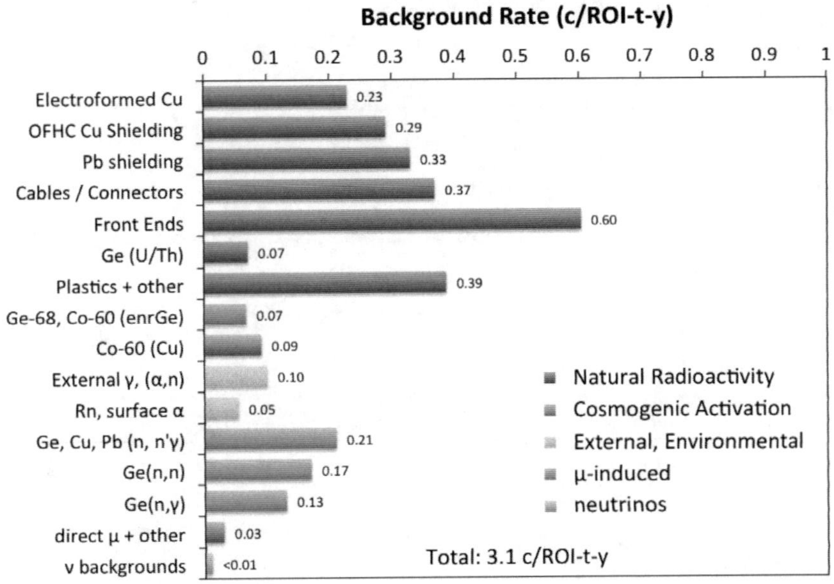

Figure 6. Estimated background contributions in the $0\nu\beta\beta$ decay ROI at MJD. The estimations come from simulation studies based on material assay results as well as existing measurements of various physics processes, and some of the estimations are upper limits. The contributions sum to $\leq$ 3.1 counts/(ROI-t-y) in the Demonstrator.

3. Summary

The assembling and construction of the Majorana Demonstrator is proceeding at the 4850' level of the Sanford Underground Research Facility. The shield is near completion and a prototype module with ten natural HPGe detectors was constructed and has been in commissioning. The first production module with 20-kg HPGe P-PC detectors, mostly enriched in ^{76}Ge will be completed and enter commissioning in early 2015; the second production module is expected to be completed in late 2015.

4. Acknowledgments
This material is based upon work supported by the U.S. Department of Energy, Office of Science, Office of Nuclear Physics. We acknowledge support from the Particle Astrophysics Program of the National Science Foundation. This research uses these US DOE Office of Science User Facilities: the National Energy Research Scientific Computing Center and the Oak Ridge Leadership Computing Facility. We acknowledge support from the Russian Foundation for Basic Research. We thank our hosts and colleagues at the Sanford Underground Research Facility for their support.

References
[1] Y. Fukuda *et al.* [Super-Kamiokande Collaboration], 1998 *Phys. Rev. Lett.* **81** 1562
[2] Q. R. Ahmad *et al.* [SNO Collaboration], 2001 *Phys. Rev. Lett.* **87** 071301
[3] K. A. Olive *et al.* [Particle Data Group], 2014 *Chin. Phys.* C **38** 090001
[4] F. T. Avignone III, S. R. Elliott, and J. Engel, 2008 *Rev. Mod. Phys.* **80** 481
[5] L. Camilleri, E. Lisi, and J. F. Wilkerson, 2008 *Annu. Rev. Nucl. Part. Sci.* **58** 343
[6] S. R. Elliott, 2012 *Mod. Phys. Lett.* A **27** 1230009
[7] N. Abgrall *et al.*, 2014 *Advances in High Energy Physics* **2014** 365432
[8] B. Schwingenheuer, 2013 *Ann. Phys. (Berlin)* **525** 4
[9] M. Goeppert-Mayer, 1935 *Phys. Rev.* **48** 512
[10] S. R. Eillott, A. A. Hahn, and M. K. Moe, 1987 *Phys. Rev. Lett.* **59** 18
[11] R. Saakyan, 2013 *Annu. Rev. Nucl. Part. Sci.* **63** 503
[12] `http://www.sanfordlab.org`; also see J. Heise, 2014 *AIP Conf. Proc.* **1604** 331
[13] G. K. Giovanetti, *et al.*, A Dark Matter Search with MALBEK *Preprint* arXiv:1407.2238
[14] P. N. Luke, F. S. Goulding, N. W. Madden, and R. H. Pehl, 1989 *IEEE Transactions on Nuclear Science* **36** 926
[15] P. S. Barbeau, J. I. Collar, and O. Tench, 2007 *Journal of Cosmology and Astroparticle Physics* **09** 009
[16] Canberra Industries, Meriden, Conn, USA, 2009.
[17] ORTEC, Oak Ridge, Tenn, USA, 2009.
[18] W. Xu *et al.*, 2014 Testing the Ge detectors for the MAJORANA DEMONSTRATOR *Preprint* arXiv:1404.7399
[19] E. W. Hoppe *et al.*, 2014 *NIM* A **764** 116.
[20] E. Aguayo *et al.*, 2013 *NIM* A **709** 17.
[21] M. A. Howe, G. A. Cox, P. J. Harvey *et al.*, 2004 *IEEE Transactions on Nuclear Science* **51** 878
[22] C. Cuesta *et al.*, 2014 Background model of the MAJORANA DEMONSTRATOR *Preprint* arXiv:1405.1370

MAJORANA Collaboration's Experience with Germanium Detectors

S. Mertens[1], N. Abgrall[1], F. T. Avignone III[2,3], A. S. Barabash[4], F. E. Bertrand[3], V. Brudanin[5], M. Busch[6,7], M. Buuck[8], D. Byram[9], A.S. Caldwell[10], Y-D. Chan[1], C. D. Christofferson[10], C. Cuesta[8], J. A. Detwiler[8], Yu. Efremenko[11], H. Ejiri[12], S. R. Elliott[13], A. Galindo-Uribarri[3], G. K. Giovanetti1[4,7], J. Goett[13], M. P. Green[3], J. Gruszko[8], I. Guinn8, V. E. Guiseppe2, R. Henning14,7, E. W. Hoppe[15], S. Howard[10], M. A. Howe[14,7], B. R. Jasinski[9], K. J. Keeter[16], M. F. Kidd[17], S. I. Konovalov[4], R. T. Kouzes[15], B. D. LaFerriere[15], J. Leon[8], J. MacMullin[14,7], R. D. Martin[9], S. J. Meijer[14,7], J. L. Orrell[15], C. O'Shaughnessy[14,7], N. R. Overman[15], A. W. P. Poon[1], D. C. Radford[3], J. Rager[14,7], K. Rielage[13], R. G. H. Robertson[8], E. Romero-Romero[11,3], M. C. Ronquest[13], B. Shanks[14,7], M. Shirchenko[5], N. Snyder[9], A. M. Suriano[10], D. Tedeschi[2], J. E. Trimble[14,7], R. L. Varner[3], S. Vasilyev[5], K. Vetter[1,18], K. Vorren[14,7], B. R. White[3], J. F. Wilkerson[14,7,3], C. Wiseman[2], W. Xu[13], E. Yakushev[5], C-H. Yu[3], and V. Yumatov[4]
The MAJORANA Collaboration

1 Nuclear Science Division, Lawrence Berkeley National Laboratory, Berkeley, CA, USA
2 Department of Physics and Astronomy, University of South Carolina, Columbia, SC, USA
3 Oak Ridge National Laboratory, Oak Ridge, TN, USA
4 Institute for Theoretical and Experimental Physics, Moscow, Russia
5 Joint Institute for Nuclear Research, Dubna, Russia
6 Department of Physics, Duke University, Durham, NC, USA
7 Triangle Universities Nuclear Laboratory, Durham, NC, USA
8 Center for Experimental Nuclear Physics and Astrophysics and Department of Physics, University of Washington, Seattle, WA, USA
9 Department of Physics, University of South Dakota, Vermillion, SD, USA
10 South Dakota School of Mines and Technology, Rapid City, SD, USA
11 Department of Physics and Astronomy, University of Tennessee, Knoxville, TN, USA 12 Research Center for Nuclear Physics and Department of Physics, Osaka University, Ibaraki, Osaka, Japan
13 Los Alamos National Laboratory, Los Alamos, NM, USA
14 Department of Physics and Astronomy, University of North Carolina, Chapel Hill, NC, USA 15 Pacific Northwest National Laboratory, Richland, WA, USA
16 Department of Physics, Black Hills State University, Spearfish, SD, USA
17 Tennessee Tech University, Cookeville, TN, USA
18 Alternate Address: Department of Nuclear Engineering, University of California, Berkeley, CA, USA

E-mail: smertens@lbl.gov

Abstract. The goal of the Majorana Demonstrator project is to search for $0\nu\beta\beta$ decay in ^{76}Ge. Of all candidate isotopes for $0\nu\beta\beta$, ^{76}Ge has some of the most favorable characteristics. Germanium detectors are a well established technology, and in searches for $0\nu\beta\beta$, the high purity germanium crystal acts simultaneously as source

Journal of Physics: Conference Series **606** (2015) 012005 doi:10.1088/1742-6596/606/1/012005

and detector. Furthermore, p-type germanium detectors provide excellent energy resolution and a specially designed point contact geometry allows for sensitive pulse shape discrimination. This paper will summarize the experiences the MAJORANA collaboration made with enriched germanium detectors manufactured by ORTEC®® .

The process from production, to characterization and integration in MAJORANA mounting structure will be described. A summary of the performance of all enriched germanium detectors will be given.

1. Introduction

A unique way to explore the nature of the neutrino is the search for neutrinoless double beta decay ($0\nu\beta\beta$) [1, 2, 3]. Observation of $0\nu\beta\beta$-decay would decisively prove that neutrinos are Majorana particles and that lepton number is violated. The MAJORANA DEMONSTRATOR will perform a search for $0\nu\beta\beta$-decay in ^{76}Ge [4, 5]. The experiment is currently under construction at the Sanford Underground Facility (SURF) in South Dakota, USA.

The MAJORANA DEMONSTRATOR will use an array of 40 kg of high purity germanium detectors, up to 30 kg of whichwill be enriched to 87% in ^{76}Ge, surrounded by passive and active shielding. The major goal is to demonstrate a path forward to achieving a background rate at or below 3 counts/(ROI-t-y) in the 4 keV region of interest (ROI) around the 2039 keV Q-value of the ^{76}Ge $0\nu\beta\beta$ decay. This is required for the next generation of tonne-scale germanium-based $0\nu\beta\beta$-decay searches that are designed to probe the neutrino mass scale in the inverted-hierarchy region.

Beyond the usage of ultra-clean material and extensive shielding, a key feature to achieve a low background level is a high energy resolution to minimize the ROI and pulse-shape discrimination capability to distinguish signal from background events. These properties are provided by High Purity (HP) P-type Point Contact (PPC) germanium (Ge) detectors. By enriching the natural germanium to 87% in ^{76}Ge the detectors can be used as source and detector simultaneously. At the time of this presentation, 30 enriched detectors with average mass of 840 g have been delivered and successfully tested at SURF. All detectors met the requirements during characterization in vendor cryostat and are now being assembled in the final MAJORANA DEMONSTRATOR experiment.

2. Production

The MAJORANA enriched ^{76}Ge material was produced by the Electrochemical Plant (ECP) in Zelenogorsk, Russia. ECP has a long history of supplying large quantities of enriched ^{76}Ge material to the world scientific community, e.g. to the GERmanium Detector Array (GERDA) Experiment [6]. A total of about 42.5 kg of enriched ^{76}Ge material has been purchased and provided for the MAJORANA DEMONSTRATOR. The germanium was isotopically enriched in ^{76}Ge from 7.8% to 87% through centrifuging germanium tetrafluoride ($^{76}GeF_4$) and then converted to germanium oxide ($^{76}GeO_2$).

All material in the form of $^{76}GeO_2$ was then shipped to Oak Ridge, TN, USA. Transportation in a special steel-shielding container reduced the activation of ^{68}Ge and ^{60}Co due to cosmic-ray exposure by factors of 10 and 15, respectively. The total effective (i.e. equivalent exposure at sea level) exposure including the shipment and enrichment is estimated to be 12.4 days. At Electrochemical Systems Inc. (ESI) in Oak Ridge, Tennessee the $^{76}GeO_2$ material was reduced via high temperature reduction in a hydrogen atmosphere. Subsequently, the material was zone- refined to electronic grade ($\sim 10^{13}$ net impurities per cubic centimeter) bars, see Fig. 1. The yield at ESI for all 42.5 kg of ^{76}Ge was > 98%.

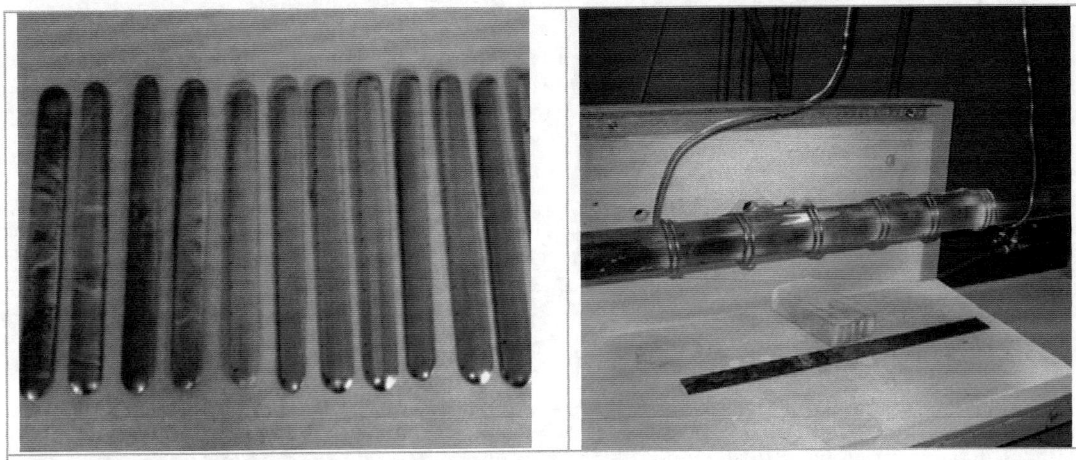

Figure 1 a) Ge metal bars after reduction to 76Ge from 76GeO2. b) Zone refinement setup at ESI

At the time of this presentation, scrap material is being recycled, to further increase the yield.

At ORTEC® [7] the manufacturing of the detectors took place after a second zone refinement to ~10^{11} net impurities per cubic centimeter. Before the shipment to SURF the first characterization campaign was performed at ORTEC® . By storing the germanium in a cave with an overburden of 40 m of rock, located about 8 km from the processing facility, a total effective exposure at ESI and ORTEC® of ~14 days was estimated. Finally, the detectors were transported to SURF by car, leading to an additional exposure time of ~35 h. Overall, the total effective sea-level equivalent exposure is approximately a factor of a few better than the project specification of no more than 100 days. At the time of this presentation, 25.2 kg of enriched germanium detectors are underground and MAJORANA is aiming for additional ~4 kg of detectors from the recovery of scrap material.

3. Characterization in Vendor Cryostats

All enriched detectors are extensively tested and characterized in their vendor cryostat both at ORTEC® and at SURF. The tests include measurements of the mass, impurity concentration (only measured at ORTEC®), depletion and operating voltages, leakage current, energy resolution, electronic noise, dead layer, relative efficiency compared to a 3x3-inch NaI detector, and pulse-shape discrimination performance. In this section the basic working principle of PPC Ge detectors with respect to these properties is outlined, the main characterization results are presented, and details of pulse-shape discrimination performance are discussed.

3.1. PPC geometry

The bulk of the detector is p-type germanium. P-type material is advantageous for efficient charge collection and therefore high energy resolution. Furthermore, it is preferable with respect to ease of handling and blocking of surface alphas. The lithiated surface of the detector provides the n^+-type contact, and a small p+ point contact is created via boron implantation. The n^+- type contact is held at positive bias voltage, whereas the point contact is kept a zero potential, see Fig. 2. The surface between the point contact and the n^+-type surface is covered with a

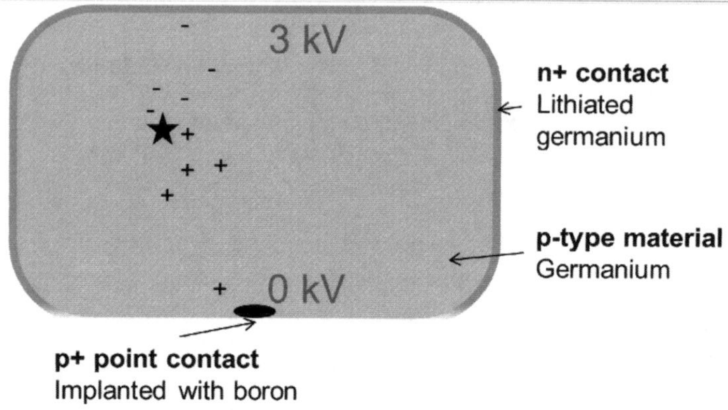

Figure 2. Schematic view of a PPC Ge detector. The green line indicates the lithiated n^+-type contact, which is held at positive bias voltage. Here 3 kV is chosen as a typical example. The small point contact is illustrated as a black oval point at the bottom of the detector, it is kept at zero potential. In this configuration the holes created by an energy deposition in the bulk of the detector move to the point contact, whereas the electrons move to the n^+-type contact.

passivation layer and is one of the most sensitive parts of the detector, any damage or contamination on this surface can lead to large leakage currents and malfunctioning of the detector.

The particular geometry of a PPC detector creates a weighting potential, such that the leading edge of the signal is largely created only when the holes arrive at the immediate vicinity of the point contact. This leads to pulse shapes which are almost independent of their point of origin, and furthermore creates distinct deviations from this shape when an event is composed of energy depositions at multiple positions in the detector. More details of the pulse shape discrimination capability of PPC detector are discussed in section 3.3.

3.2. General performance results

In order to maximize the exposure and keep the number of channels at a manageable level, large mass detectors are preferable. At the same time, in order to fully deplete the detectors at a technically feasible operating voltage, the impurity concentration must be kept at a low level. With an average mass of 840 g and an average impurity concentration of $7 \cdot 10^9$ net impurities per cubic centimeter all MAJORANA enriched detectors are operated at a bias voltage of less than 5 kV. Furthermore, the average leakage current of all detectors is only ~11 pA at full bias voltage.

A major advantage of P-type point contact detectors is their excellent energy resolution, which minimizes the energy width of the ROI, and maximizes the signal-to-background ratio. The resolution is typically measured via the FWHM of the 1.3 MeV line of a 60Co calibration source. With an average FWHM of 1.9 keV at 1.3 MeV, all 30 detectors exceed the vendor specification of 2.3 keV, see Fig. 3. ORTEC® and SURF measurements are in very good agreement. At the 122 keV line of 57Co an average FWHM of 624 eV is achieved at ORTEC® .

A low electronic noise level allows for a low energy threshold. This is preferable for example to assure

a reliable

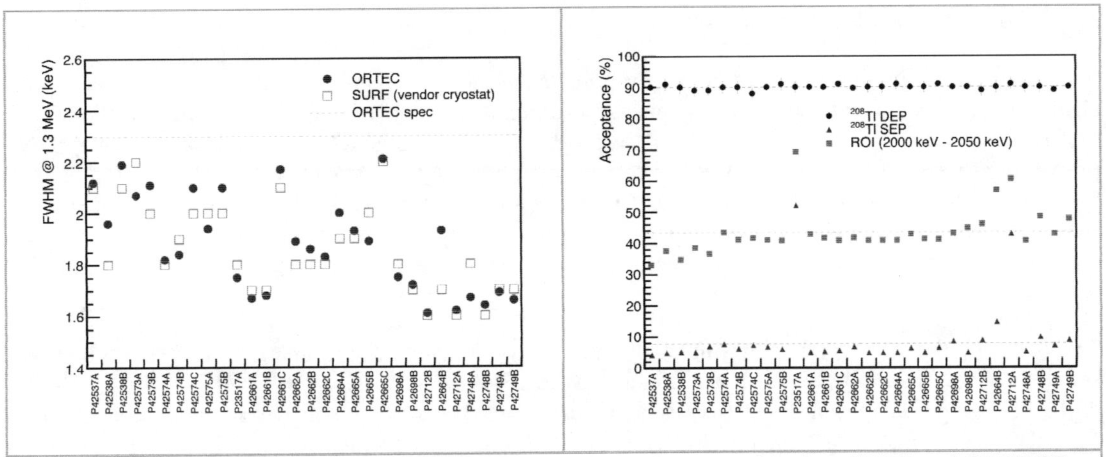

Figure 3. (a): Energy resolution of all 30 enriched detectors, measured both at SURF and ORTEC® . All detectors meet the specification. (b): Pulse shape discrimination performance of all 30 enriched detectors. By fixing the acceptance in the double-escape peak (DEP) to 90%, the counts in the single-escape peak can be reduced to less than 10% for all detectors but three. All detector will still be used in the final MAJORANA DEMONSTRATOR experiment. The degraded performance of P23517A can likely be related to its very low impurity concentration, see Section 3.4. A flat edge of the crystal might explain the pulse-shape discrimination behavior of P42664B and the performance of P42712A can be explained by an instable depletion voltage during the measurement.

rejection of background due to cosmogenically induced ^{68}Ge. This background is expected to be reduced to 0.14 counts/ROI/t/y averaged over the first year, by applying an single-site time correlation cut for five ^{68}Ga-half-lives following 10-keV K-shell deexcitations. If the 1-keV L-shell de-excitations can also be tagged, this background may be reduced by another order-of-magnitude.

Furthermore, a low energy threshold allows for the option of extending the physics reach of MAJORANA to a search for new physics, such as low-mass WIMPS [8], which would leave a

signal at the low energy end of the spectrum. The electronic noise is determined at ORTEC® via the FWHM of a pulser signal. All detectors meet the requirement of $FWHM_{pulser} < 500$ eV. To maximize the active volume of the detectors a thin dead layer is required. On the other hand, a minimal thickness is advantageous [9], as it prevents surface alphas to deposit energy in the ROI. With an average dead layer of 0.8 mm all 30 enriched detectors meet the requirements.

The dead layer has been determined by comparing the ratio of the peak heights of 81 keV and 356 keV ^{133}Ba lines to simulation to determine the relative difference in attenuation due to the inactive Ge. Measurements at ORTEC® and SURF are in good agreement.

The relative efficiency compared to a 3 × 3 inch NaI(Tl) detector is determined with a ^{60}Co source of known strength, which is placed at a given distance to the detector, on axis with the center of the crystal. The counts in the photo peak at 1.33 MeV are measured and then normalized to the efficiency of a standard 3x3-inch NaI detector. The measured efficiencies agree with what ORTEC® measured and our simulation. Measurements at ORTEC® and SURF are in good agreement.

Journal of Physics: Conference Series **606** (2015) 012005 doi:10.1088/1742-6596/606/1/012005

3.3. Pulse shape discrimination performance

A key feature of the point contact geometry is its capability of pulse-shape discrimination. The two electrons emitted in $0\nu\beta\beta$ deposit their energy in a small volume of the detector and are therefore referred to as single-site event. A large class of background events is due to gammas which have high probability to deposit their energy at multiple positions in the detectors, and are therefore called multi-site events. The point-contact geometry creates a weighting potential in the detector such that the pulse shape of single-site events is significantly different from multi- site events, see Fig. 4. A simple discriminative property of the pulse shape is the ratio of the maximum current (A) and the amplitude (E) of the pulse.

To determine the effectiveness of single-site and multi-site event discrimination, a ^{228}Th calibration source is used. In the decay of ^{208}Tl a high energy (2.615 MeV) gamma is produced, which has a high probability of producing an e+ e− pair in the Ge detector. The subsequent two gammas produced by annihilation of the positron can either fully deposit their energy in the detector (full energy peak), only one gamma deposits energy in the detector (single escape peak), or both gammas leave the detector without interaction (double escape peak). The latter generates a typical single site event, whereas the single escape peak is populated with multi- site events. These two event classes are ideal to evaluate the efficiency of the pulse shape discrimination.

By fixing the acceptance of single-site events to 90%, the cut efficiency for the single-escape peak and the ROI is determined. For all enriched detectors, with the exception of one, see Fig. 3, the single-escape events could be reduced to less than 10%. The number of events in the ROI is reduced to about 40%, which corresponds to the expected number of single-site events in this energy range.

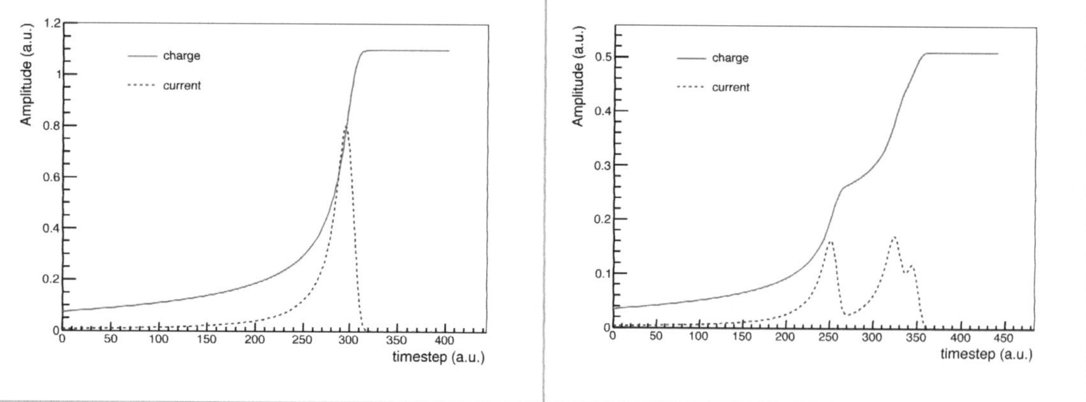

Figure 4. (a): Simulated single-site event. (b): Simulated multiple-site event. The red solid line shows the charge as as a function of time, the blue dotted line corresponds to its derivative, i.e. the current. The 10% to 90% rise time of the signal is created only while the charge is in very close vicinity to the point contact, see Fig. 5. Consequently, the ratio of maximum current to the amplitude of the signal (energy) is very similar for all single-site events and different for multiple-site events.

3.4. Impact of ultra-low impurity concentrations on pulse-shape discrimination performance

Detector P23517A which showed the poorest pulse-shape discrimination performance (see Fig. 3) also exhibits the lowest depletion voltage, which can be related to an extremely low impurity concentration. Dedicated simulations with the simulation software *siggen* [10] reveal a clear impact of ultra-low impurity concentrations on the pulse-shape discrimination performance.

This fact can be understood by considering the electric field in a detector. The total electric potential is given as a superposition of the potential created by the bias voltage and the potential created by the impurity concentration. In particular, the field in the corners is governed by the impurity concentration, as can be seen in Fig. 5. To first order, the time difference $\Delta t \approx \Delta x / v_{init}$ between the charges arriving at the point contact is given by their initial spatial separation Δx relative to their initial velocity v_{init}. As the velocity is determined by the electric field, Δt and correspondingly the maximum current A, is small for charges created in the corners of an ultra-low impurity.

The spatial separation is driven by diffusion and self-repulsion. Analytic estimates show that, while charges move in the low field region (in low-impurity detectors the field can drop below 50 V/cm in the corners), these effects can cause a spread of the charges of up to about 1 mm. As a first approximation, the spatial separation of charges is implemented in the simulation code by a convolution of the simulated signal by a Gaussian with a width corresponding to the charge cloud size.

To validate the model a set of dedicated tests with a natural ultra-low purity detector with approximately $0.5 \cdot 10^9$ net impurities per cubic centimeter has been performed. First, the radial dependence of the cut parameter A/E was determined with the help of a radial scan using a ^{241}Am source. Secondly, a coincidence measurement was performed to determine the drift time of the charges from the point of their creation t_{init} to their arrival time t_{final} at the point contact. To this end, a collimated ^{22}Na source, which emits two back-to-back gammas, was placed between the Ge detector and a NaI detector. The latter provides t_{init} and the PPC detector determines the arrival time t_{final}. With this information the behavior of A/E as a function of the drift time was investigated.

Both studies confirm that the maximum current A is reduced for charges created in the low-field corners of an ultra-low impurity detector, see Fig. 6. This position dependence of the cut parameter is the reason for the degraded pulse-shape discrimination performance of ultra-low impurity detectors. Without the information on the position or the drift time (neither of which are provided by a non-segmented point-contact detector) the A/E cut removes events based on the creation position even if they are single-site in nature.

4. Integration in MAJORANA mount

After completion of the characterization in the vendor cryostat, the detectors are removed from the vendor cryostat and mounted in MAJORANA's low-background mount. The detector mounting structure is made from ultra-pure electro-formed copper and the front-end electronics [13], which is placed in close vicinity to the detector, is made from carefully selected low-background material and with a

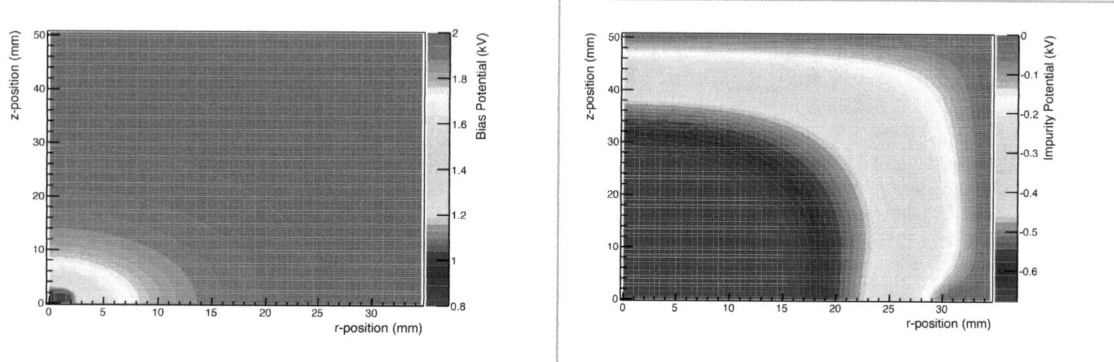

Figure 5. (a): Electric potential created by the bias voltage. In this example the bias is voltage is 3 kV. This potential determines the weighting potential of the detector. (b): Electric potential created by the impurity concentration. At the conductive surfaces of the detector the potential is forced to zero volts, whereas the high resistive passivation layer can be at a finite potential. In this example model of a detector with an impurity concentration of $3.2 \cdot 10^9$ net impurities per cubic centimeter at the top of the detector and an impurity gradient of $0.25 \cdot 10^9$ change in impurity density per centimeter was assumed. The total electric potential in the detector is given by the sum of the potentials in a) and b).

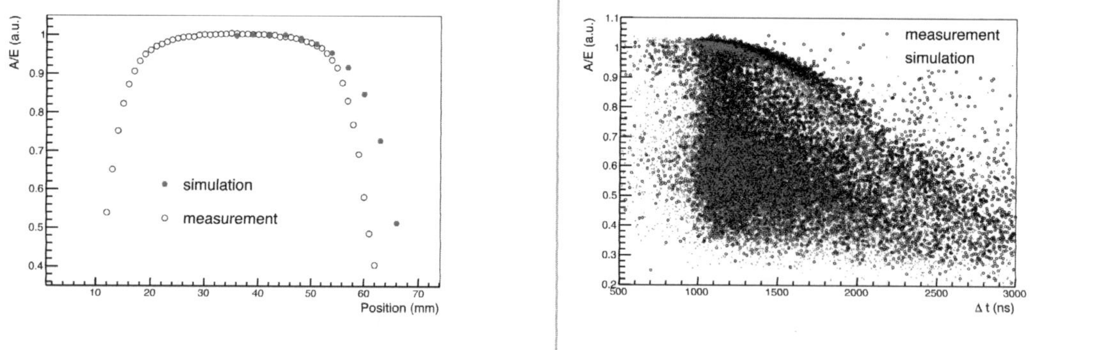

Figure 6. (a): Comparison of measurement and simulation for an ^{241}Am radial scan across the top of a ultra-low purity detector. Both measurement and simulation show a clear decrease of A/E at larger radii, corresponding to charges being created in the low-field corners of the detector. (b): Comparison of measurement and simulation of a coincidence measurement with a ^{22}Na source. Again, the simulation reproduces the decreased A/E for large drift times Δt, corresponding to charges created in the low-field regions. A constant cut in A/E would cut events at large drift times and large radii and thereby degrade the pulse-shape discrimination performance. For these simulations a detector with an impurity concentration of $1.6 \cdot 10^9$ net impurities per cubic centimeter at the top of the detector and an impurity gradient of $0.39 \cdot 10^9$ change in impurity density per centimeter was assumed.

minimal mass. Figure 7 shows the integration of the detectors in the MAJORANA mounts. To closely pack the detectors in the final detector array, a combination of 4–5 natural and enriched detectors are stacked into a so-called detector string. Seven of these strings are placed in a module cryostat, two of which will be used to house and operate the MAJORANA detectors [4]. The natural detectors used by MAJORANA are Broad Energy Ge (BEGe) detectors, produced by Canberra [11]. Their particular properties and performances are not detailed in this proceedings.

Prior to the final installation of the strings in the module cryostats, they are again tested in the a so-called String Test Cryostat (STC). The goal of these tests is on the one hand to check the integrity of the detector, front-end electronics and HV connection after the re-installation of the detector in the MAJORANA mount, and on the other hand, this configuration allows the measurement of the crystal axis orientation of each detector in the string and the homogeneity of the dead layer along the side of each detector. The crystal axis orientation is of particular interest in view of a possible axion search with MAJORANA [12].

Fig. 8 shows a calibration measurement with a ^{133}Ba and ^{60}Co source taken with a detector operated in a STC. This measurement shows that the detector is operational at its standard operating voltage, i.e. the passivation layer has not been harmed in the installation process, the front-end electronics is intact, and the HV connection is functional. In this more complex setup (compared to the vendor cryostat) an energy resolution of FWHM ≈ 2 keV at the 1.3 MeV ^{60}Co-line could be achieved, which is similar to what was measured in the vendor cryostat and meets the MAJORANA requirements.

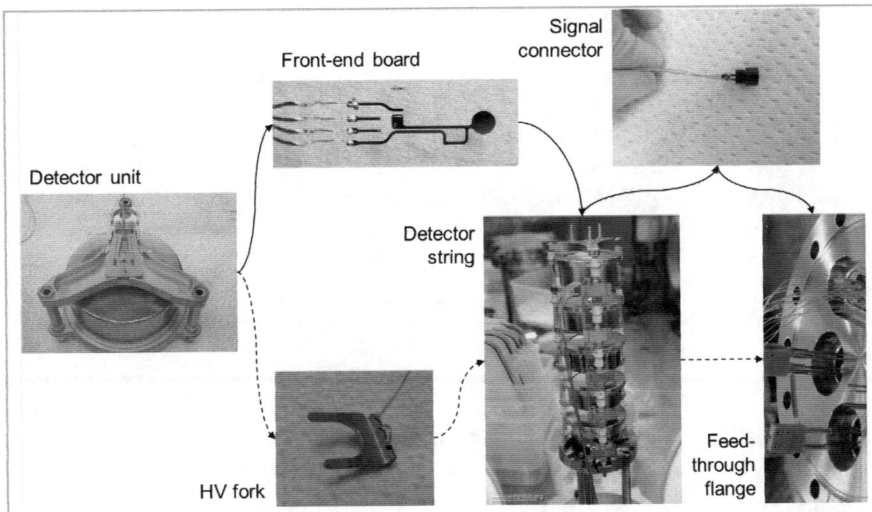

Figure 7. Mounting of detectors in MAJORANA mount. The left photograph in the diagram shows a natural detector mounted in its electro-formed copper mount. The signal is extracted using a low mass front-end board (LMFE) [13], HV is applied via the HV fork which is clamped between a PTFE nut and the HV ring at the bottom of the detector. 4–5 detector units are stacked to form a detector string. The HV line is directly brought to the feedthrough flange, whereas the signal line passes through an ultra-clean connector on top of the coldplate beneath which the string is attached.

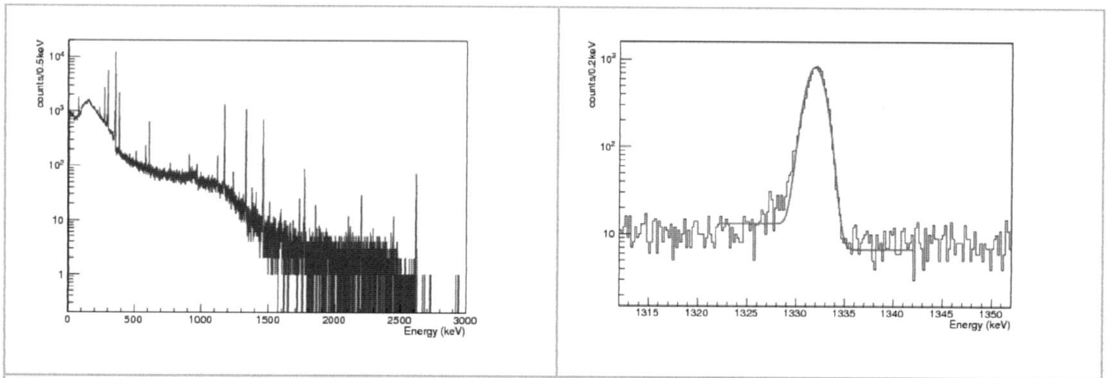

Figure 8. (a): Calibration spectrum of a detector in a string test cryostat. (b): Zoom into the 1.33 MeV line of the ^{60}Co calibration source. The energy resolution for this peak is FWHM = 1.9 keV. The small low energy tail is expected to be due to fact that this data was recorded without pole-zero correction.

5. Conclusion

MAJORANA experienced a good working relationship with ECP in Russia, ESI, and ORTEC® in the USA. The total sea-level exposure time for the enriched detectors could be kept better than the project specification of no more than 100 days. After reprocessing of scrap material MAJORANA expects ~ 30 kg of enriched detectors. All enriched detectors met requirements during characterization in the vendor cryostat at ORTEC® and SURF. Dedicated simulations and measurements indicate that the slightly degraded pulse-shape discrimination behavior of one of the detectors is likely due to its ultra-low purity level. At the time of this presentation MAJORANA is in the process of installing the detectors in specially designed low-background detector mounts and strings. First measurements in the string test cryostats show positive results with respect to integrity and performance of the detectors in the MAJORANA mounts.

References

[1] M. Goeppert-Mayer, Double Beta-Disintegration, Phys. Rev. 48, 512 (1935)
[2] E. Majorana, Teoria simmetrica dellelettrone e del positrone, Nuovo Cimento 14, 171 (1937)
[3] F. T. Avignone III, S. R. Elliott, and J. Engel Double beta decay, Majorana neutrinos, and neutrino mass, Rev. Mod. Phys. 80, 481, (2008)
[4] N. Abgrall et al. (MAJORANA Collaboration), The MAJORANA DEMONSTRATOR Neutrinoless Double Beta Decay Experiment, Adv. High Energy Physics 2014, Article ID 365432, (2014)
[5] W. Xu et al. (MAJORANA Collaboration), The MAJORANA DEMONSTRATOR: A Search for Neutrinoless Double- beta Decay of 76Ge , to be published in Journal of Physics: Conference Series
[6] K. H. Ackermann et al. (GERDA Collaboration), The GERDA experiment for the search of $0\nu\beta\beta$ decay in 76Ge, European Physical Journal C 2013 73:2330, (2013)
[7] ORTEC® , Advanced Measurement Technology, Inc, 801 South Illinois Avenue, Oak Ridge, Tennessee 37831- 0895, United States
[8] G. K. Giovanetti et al. (MAJORANA Collaboration), A Dark Matter Search with MALBEK,

submitted to Proceedings of the 13th International Conference on Topics in Astroparticle and Underground Physics, TAUP 2013

[9] B. J. Brennan and R. G. Lyons, Ranges of alpha particles in various media Ancient TL, Volume 7, No 2, (1989)

[10] D. Radford et al., Pulse shape simulation for Ge detectors, to be published

[11] Canberra Industries, Inc., 800 Research Parkway, Meriden, CT 06450, U.S.A.

[12] R. J. Creswick et al., Theory for the direct detection of solar axions by coherent Primakoff conversion in germanium detectors, Phys Lett B 427 (1988) 235

[13] N. Abgrall et al. (MAJORANA Collaboration), The MAJORANA low-noise low-background front-end electronics, submitted to Proceedings of the 13th International Conference on Topics in Astroparticle and Underground Physics, TAUP 2013

Results on neutrinoless double beta decay of ^{76}Ge from GERDA Phase I

Dimitrios Palioselitis
on behalf of the GERDA collaboration
Max-Planck-Institut für Physik, München, Germany
E-mail: dimitris@mpp.mpg.de

Abstract. The Germanium Detector Array (GERDA) experiment is searching for the neutrinoless double beta ($0\nu\beta\beta$) decay of ^{76}Ge by operating bare germanium diodes in liquid argon. GERDA is located at the Gran Sasso National Laboratory (LNGS) in Italy. During Phase I, a total exposure of 21.6 kg yr and a background index of 0.01 cts/(keV kg yr) were reached. No signal was observed and a lower limit of $T^{0\nu}_{1/2} > 2.1 \cdot 10^{25}$ yr (90% C.L.) is derived for the half life of the $0\nu\beta\beta$ decay of ^{76}Ge.

1. Introduction

Neutrino accompanied double beta ($2\nu\beta\beta$) decay is a second order weak process predicted by the Standard Model of particle physics. This decay has been observed for several isotopes and the experimentally determined half-lives lie in the range of 10^{19} to 10^{24} years [1, 2].

Neutrinoless double beta ($0\nu\beta\beta$) decay is a process that violates lepton number conservation by two units, and its observation would indicate physics beyond the Standard Model [3, 4, 5]. Furthermore, it would prove that neutrinos have a Majorana mass component. This process has not been observed so far and the half-life limits set on $0\nu\beta\beta$ decay for ^{76}Ge lie in the range of $(1.6 - 1.9) \times 10^{25}$ years [6, 7, 8]. In 2004, part of the HdM collaboration claimed an observation of $0\nu\beta\beta$ decay [9], reporting a half-life of $T^{0\nu}_{1/2} = (1.19^{+0.37}_{-0.23}) \times 10^{25}$ years. The experimental signature of $0\nu\beta\beta$ decay is a monoenergetic peak of the sum electron kinetic energy at the Q-value of the decay, $Q_{\beta\beta} = 2039\,\mathrm{keV}$, above the continuous energy spectrum of the $2\nu\beta\beta$ decay.

The GERDA experiment is introduced in section 2. The results from GERDA Phase I are summarised in the following sections. The measurement of the half-life of $2\nu\beta\beta$ decay, the modelling of the background energy spectrum and the background discrimination methods are discussed in sections 3, 4 and 5, respectively. The result on the search of $0\nu\beta\beta$ decay is presented in section 6. The status of the ongoing transition to Phase II is presented in section 7.

2. The GERDA experiment

The GERDA experiment searches for the $0\nu\beta\beta$ decay of ^{76}Ge [10]. The experiment is located at the Laboratori Nazionali del Gran Sasso (LNGS) in Italy. GERDA operates bare germanium diodes inside liquid argon (LAr) which serves as a coolant and as shielding. The array of germanium detectors is suspended inside a stainless steel cryostat filled with $64\,\mathrm{m}^3$ of LAr. The cryostat is located inside a water tank that contains $590\,\mathrm{m}^3$ of high purity water, moderating

Figure 1. An artist's view of the GERDA detector. The array of germanium detectors (1), the LAr cryostat (2), the internal copper shield (3), the water tank (4), the clean room (5) and the lock system (6) are indicated. Taken from [10].

ambient neutrons and gamma radiation. The water tank is instrumented with 66 photomultiplier tubes (PMTs) and operates as a Čerenkov muon veto reducing the cosmic induced background index to less than 10^{-3}cts/(keV kg yr) [11]. The rate of events coincident between the germanium detectors and the GERDA muon veto system is $(9.3 \pm 0.4) \times 10^{-5}$/s [12]. The detector strings are lowered into the cryostat from the clean room, located above the GERDA tank. An artist's view of the experimental setup is shown in figure 1.

Two types of detectors were used during GERDA Phase I. Eight p-type high purity germanium (HPGe) semi-coaxial detectors from the HdM [6] and IGEX [7] experiments were refurbished and used as the main GERDA Phase I detectors. They have an n+ conductive lithium layer and a boron implanted p+ contact, separated by a groove. They are enriched to ~86% in ^{76}Ge and have a total mass of 17 kg. Additionally, 30 enriched p-type broad energy germanium (BEGe) detectors were produced and will be used in Phase II of the experiment [13, 14]. Five of them were already deployed in GERDA during Phase I. Their total mass is 3.6 kg.

The data collected during Phase I, from November 2011 until May 2013, correspond to 492 days and a total exposure of 21.6 kg yr. The average duty cycle is 88%. The data were divided into three datasets. The *golden coaxial dataset* with an exposure of 17.9 kg yr contains all data taken with the enriched semi-coaxial detectors with the exception of a short period of approximately 30 days. This was due to increased activity after the insertion of the five BEGe detectors. This dataset, referred to as *silver coaxial dataset*, corresponds to 1.3 kg yr. The *BEGe dataset* consists of data taken with the BEGe detectors and has an exposure of 2.4 kg yr.

Regular calibration runs were taken on a weekly basis, using a ^{228}Th source, in order to determine the energy scale of the individual detectors. The energy shift between successive calibrations is less than 1 keV at $Q_{\beta\beta}$. This is due to gain drifts of the readout chain [10]. The mean exposure-weighted energy resolutions for the GERDA detectors are 4.8 ± 0.2 keV for the semi-coaxial detectors and 3.2 ± 0.2 keV for the BEGe detectors.

3. Measurement of the $2\nu\beta\beta$ decay half-life

The measurement of the half-life of the neutrino accompanied double beta decay of ^{76}Ge by GERDA corresponds to an exposure of 5.04 kg yr [15]. The observed energy spectrum between 600 and 1800 keV is dominated by the neutrino accompanied double beta decay of ^{76}Ge. The signal-to-background ratio in this energy range is on average 4:1. A global model was fitted to the observed energy spectra above the cosmogenic ^{39}Ar background, which dominates the energy spectrum below 565 keV. The model contains the $2\nu\beta\beta$ decay of ^{76}Ge and three independent background contributions from ^{42}K, uniformly distributed in liquid argon, as well as ^{214}Bi and

Journal of Physics: Conference Series **606** (2015) 012006 doi:10.1088/1742-6596/606/1/012006

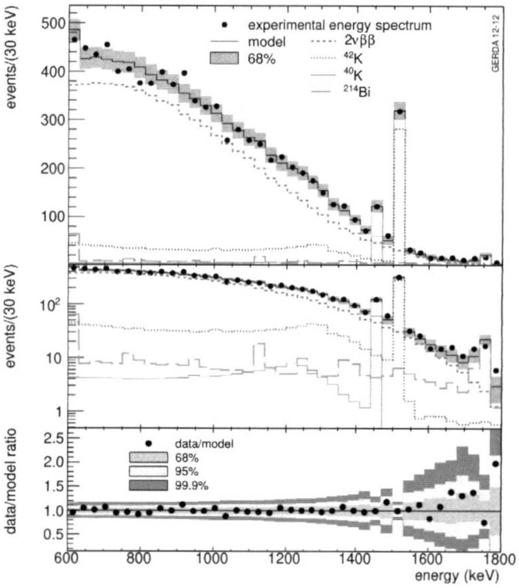

Figure 2. Upper and middle panels: experimental data (markers) and the best fit model (black histogram) in linear and logarithmic scale. Individual components are shown with coloured histograms. The shaded band covers the 68% probability range for the data, calculated from the expected counts of the best fit model. Lower panel: ratio between the experimental data and the best fit model, shown with the smallest intervals containing 68%, 95% and 99.9% probability for the ratio assuming the best fit parameters. Taken from [15].

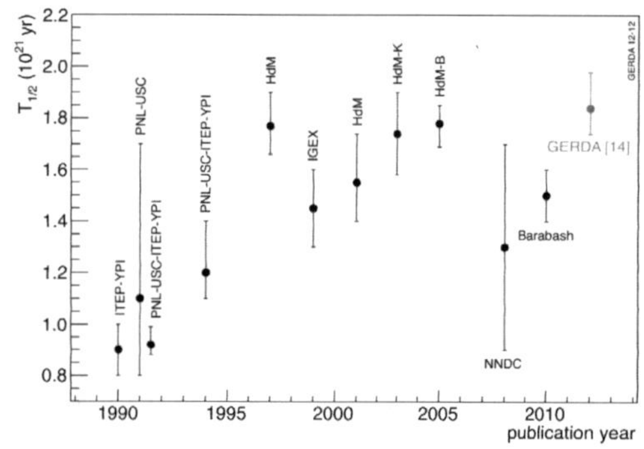

Figure 3. Experimental results for the half-life of the neutrino accompanied double beta decay, $T_{1/2}^{2\nu}$, of ^{76}Ge versus publication year. Taken from [15].

^{40}K from close sources. The presence of these sources is established by the observation of their characteristic gamma lines. Possible contributions from other background components were included in the systematic uncertainties. The spectral fit has 32 free parameters, the $2\nu\beta\beta$ half-life, the detector masses and enrichment fractions and the background contributions. The experimental energy spectrum together with the best fit model and the individual spectral contributions are shown in figure 2. The ratio between experimental data and the prediction of the best fit model is shown in the lower panel. The green, yellow and red regions are the smallest intervals containing 68%, 95% and 99.9% probability for the ratio, respectively, assuming the best fit parameters.

After marginalising over all nuisance parameters, the best half-life estimate is $T_{1/2}^{2\nu} = (1.84^{+0.14}_{-0.10}) \times 10^{21}$ yr. This result is shown in figure 3, together with previous publications and two weighted averages. The value reported by GERDA is longer than the previous measurements. There is a tendency towards higher values for more recent measurements. This is probably related to the improved signal-to-background ratio, which reduces the relevance of background

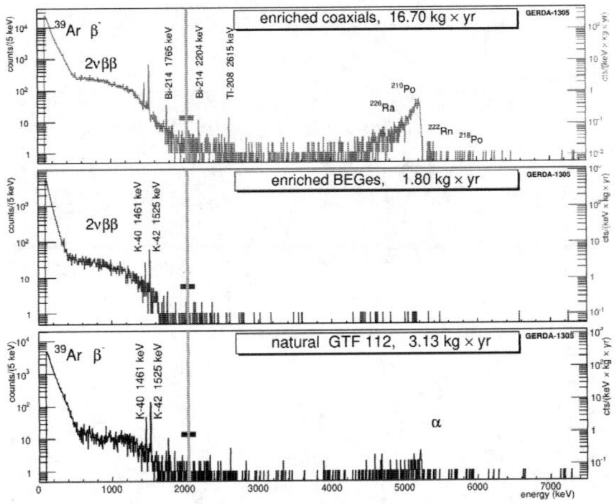

Figure 4. Energy spectra taken with all the enriched semi-coaxial (top), BEGe (middle) and non-enriched (bottom) detectors. The green band indicates the blinded region of 40 keV around $Q_{\beta\beta}$. The bars in the colour of the histogram around the blinded region indicate the 200 keV window used to determine the background index of each dataset. Taken from [12].

modelling and subtraction.

4. Background modelling for GERDA Phase I

A good understanding of the background is important in order to extract a possible $0\nu\beta\beta$ signal or to obtain a limit on the half-life of the process in case no signal events are observed. A background model was developed, prior to the $0\nu\beta\beta$ analysis, to describe the observed energy spectrum using data corresponding to an exposure of 18.5 kg yr [12]. The model contains several contributions that are either expected after material screening or established through the observation of characteristic structures in the energy spectrum. The energy spectra for the enriched semi-coaxial detectors, the BEGe detectors and one non-enriched detector are shown in figure 4. The low energy part, up to 565 keV, is dominated by the beta decay of cosmogenic ^{39}Ar. Between 600 and 1500 keV the spectra of the enriched detectors are dominated by the $2\nu\beta\beta$ decay of ^{76}Ge. Gamma lines from the decays of ^{40}K and ^{42}K can be identified in all spectra. Gamma lines from ^{60}Co, ^{208}Tl, ^{214}Bi, ^{214}Pb and ^{228}Ac are visible in the spectra of the enriched semi-coaxial detectors. A peak-like structure around 5.3 MeV in the spectrum of the enriched semi-coaxial detectors can be attributed to the decay of ^{210}Po on the detector p+ surfaces. Further peak-like structures at energies of 4.7, 5.4 and 5.9 MeV can be attributed to the alpha decays of ^{226}Ra, ^{222}Rn and ^{218}Po on the detector p+ surface, respectively. A 40 keV window around $Q_{\beta\beta}$ was kept blinded during the analysis.

The background model was obtained by fitting the simulated spectra of different contributions to the measured energy spectrum using a Bayesian approach. The high energy part of the spectrum between 3.5 and 7.5 MeV, above the Q-value of ^{42}K, was analysed first, providing a best fit for the alpha induced spectrum. This result was used along with other contributions to establish a model covering the energy range from 570 to 7500 keV. The main contributions at $Q_{\beta\beta}$ come from ^{42}K (uniform in LAr), ^{60}Co (in germanium and on the detector assembly), ^{214}Bi (on the detector assembly and p+ surface), ^{208}Tl (on the detector assembly), as well as alpha events from surface contamination and ^{222}Rn in LAr. Figure 5 shows the best fit model in black, together with the observed counts and the individual background contributions considered in the high energy alpha fit, for the *golden coaxial dataset.* Figure 6 shows the best fit model in black, together with the observed counts and the individual background contributions considered in the global fit, for the *golden coaxial dataset.* In the lower panels, the ratios of data and model are shown together with the smallest intervals of 68%, 95% and 99.9% probability for the model expectation.

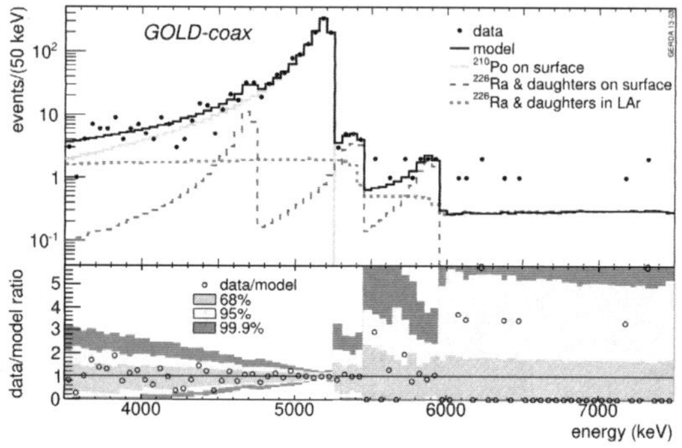

Figure 5. Experimental data (markers) and the best alpha model fit (black line) for the *golden coaxial dataset.* The lower panel shows the ratio between data and model, as well as the 68%, 95% and 99.9% probabilities for the ratios assuming the best fit parameters. Taken from [12].

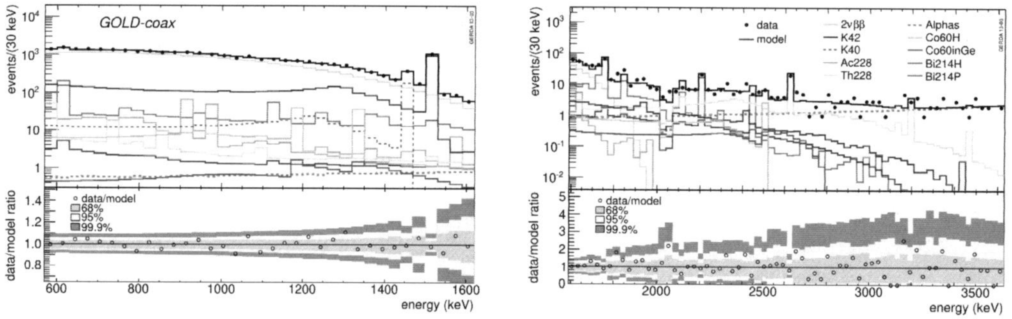

Figure 6. Experimental data (markers) and the best fit minimum model (black line) for the *golden coaxial dataset.* Individual background contributions are shown separately. In the legend, "inGe", "H" and "P", refer to contributions from the germanium detectors, the holders and the p+ surface, respectively. The lower panels show the ratio between data and model, as well as the 68%, 95% and 99.9% probabilities for the ratios assuming the best fit parameters. Taken from [12].

Two models were constructed for the background description, a minimal model consisting of well-motivated contributions (see figure 6) and a maximum model consisting of various additional contributions. They are shown in figure 7 together with the experimentally observed counts for the *golden coaxial dataset.* Both models show a good agreement with the data and their p-values do not favour one model over the other.

A partial unblinding was performed in order to confirm the validity of the background model. A 10 keV window for the semi-coaxial detectors and an 8 keV window for the BEGe detectors around $Q_{\beta\beta}$ remained blinded for the $0\nu\beta\beta$ analysis. The partially unblinded data are shown in light grey in figure 7. The agreement of the model with the data after the partial unblinding is satisfactory. It is important to note that the spectrum between 1930 and 2190 keV can be modelled with a flat background, shown in the lower panels of figure 7. The background index at the region of interest is $(17.6 - 23.8) \times 10^{-3}$ cts/keV kg yr before applying any pulse shape discrimination algorithms.

5. Pulse shape discrimination analysis

The experimental sensitivity can be improved by analysing the pulse shapes of the detector signals with the aim of rejecting background events. Pulse shape discrimination (PSD) is therefore used to separate single-site (SSE) from multi-site (MSE) events. The signature of

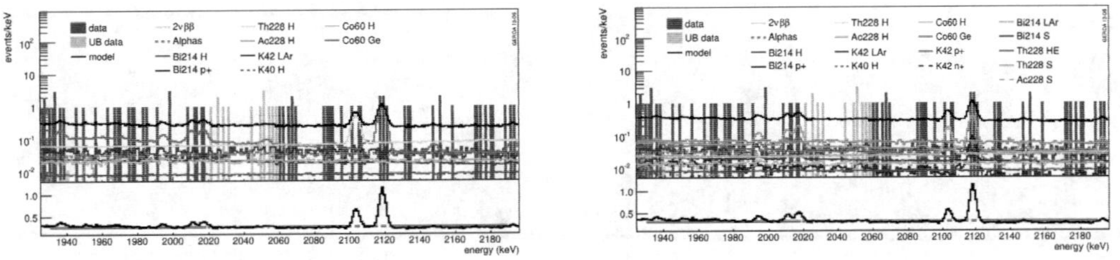

Figure 7. Experimental spectrum with the minimum (left) and maximum (right) background model around $Q_{\beta\beta}$ for the *golden coaxial dataset*. The upper panels show the experimental data (grey histograms) with the individual background contributions considered in the fit (coloured histograms). The light grey histogram corresponds to the partially unblinded data, not used for the modelling of the background. The lower panels show the best fit models fitted with a constant. In the legend, the following abbreviations are used for the location of the background contributions: H: detector holders, p+: p+ contact, LAr: uniform in liquid argon, Ge: inside the germanium crystal, n+: n+ surface, S: radon shroud, HE: heat exchanger. Taken from [12].

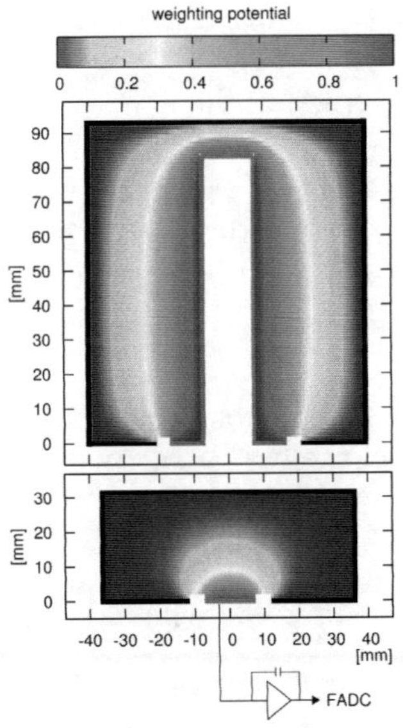

Figure 8. Cross section of a semi-coaxial (top) and a BEGe (bottom) detector. The n+ electrode is drawn in black and the p+ electrode in grey. The weighting potential inside the detector crystals is indicated through a colour map. Taken from [16].

a double beta decay is a SSE, i.e. the energy is deposited in a single location in the detector. On the other hand, MSEs, e.g. from multiple Compton scattering, deposit energy in well separated locations in the detector.

Different PSD techniques were used for the semi-coaxial and the BEGe detectors [16]. This is due to the different geometries and, hence, different electric field distributions of the detectors. The cross sections of a semi-coaxial and a BEGe detector, along with the corresponding weighting potentials, are shown in figure 8. For the semi-coaxial detectors a neural network approach was utilised, where the rising part of the charge pulse was used for the network analysis. For the BEGe detectors a mono-parametric A/E method was implemented, where A corresponds to the maximum of the current pulse and E is the reconstructed energy. For MSEs, the current pulses of the charges from different locations will have different drift times and, hence, more time-separated current pulses. Therefore, for the same total energy, E, the maximum amplitude,

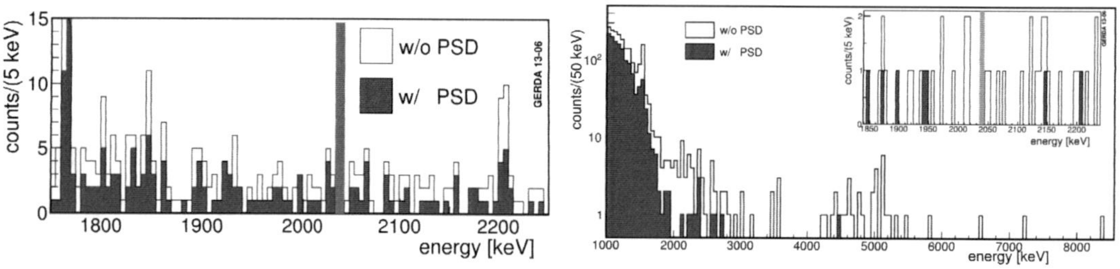

Figure 9. Energy spectrum of the semi-coaxial (left) and BEGe (right) detectors with (filled histogram) and without (open histogram) PSD selection. The blinded window is indicated with a green band. Taken from [16].

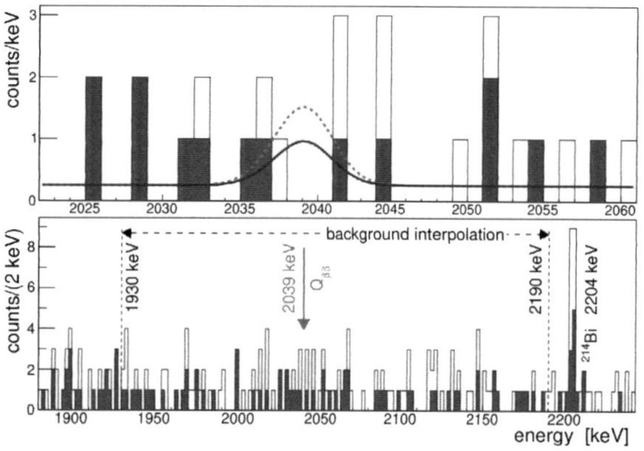

Figure 10. Energy spectrum around $Q_{\beta\beta}$ for all enriched germanium detectors with (filled) and without (open) PSD. The red dashed line corresponds to the expectation from [9] and the blue solid line is the 90% C.L. upper limit derived from the GERDA $0\nu\beta\beta$ analysis. The lower panel shows the energy region used for the background interpolation. Taken from [17].

A, will be smaller for MSEs. As a proxy of SSEs, events from the double-escape peak (DEP) at 1593 keV from the 2614 keV line of ^{208}Tl are used. Events in the full energy line of ^{212}Bi at 1621 keV are mostly MSEs and are used as the background sample. More information on pulse shapes from semi-coaxial and BEGe detectors and about the details of the analysis can be found in [16].

Figure 9 shows the result of the PSD methods applied to data for the semi-coaxial and the BEGe detectors. The events surviving the PSD selection are shown in grey. The neural network method has a $0\nu\beta\beta$ acceptance of 90% while it rejects approximately half of the background around $Q_{\beta\beta}$. The A/E method has an efficiency of 92% and rejects 80% of the background events around $Q_{\beta\beta}$. On $2\nu\beta\beta$ events, the methods have an efficiency of 85% and 91% for the semi-coaxial and BEGe detectors, respectively.

6. Results on neutrinoless double beta decay of ^{76}Ge

The combined energy spectrum from all enriched germanium detectors around the region of interest after unblinding is shown in figure 10, before (open histogram) and after (filled histogram) PSD selection. The energy region used for the background interpolation is shown in the lower panel.

After opening the blinded window, no excess of events was found above the expected background. Two analyses were performed to derive the lower limit for the half-life of $0\nu\beta\beta$ of ^{76}Ge. The baseline analysis was a frequentist analysis, where a profile likelihood fit was performed to the datasets using a common half-life. The fit function was the sum of a constant term for the background and a gaussian term for the signal. The best fit corresponded to

Journal of Physics: Conference Series **606** (2015) 012006 doi:10.1088/1742-6596/606/1/012006

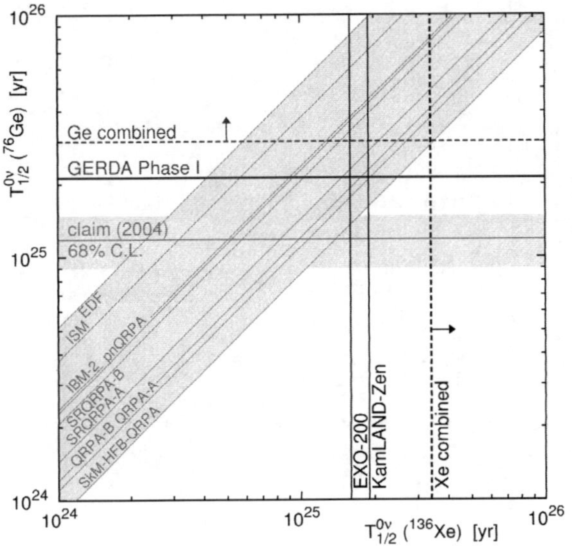

Figure 11. Limits (90% C.L.) on $T^{0\nu}_{1/2}$ of ^{76}Ge and ^{136}Xe [20, 21] compared to the signal claim for ^{76}Ge. The shaded grey bands are the predictions for the correlation of the half-lives for the two isotopes according to different nuclear matrix elements calculations. Taken from [17].

zero counts and an upper limit of 3.5 counts. The derived lower limit for the half-life of $0\nu\beta\beta$ is $T^{0\nu}_{1/2} > 2.1 \times 10^{25}$ yr at 90% confidence level, including the systematic uncertainty. The corresponding median sensitivity for the 90% C.L. limit is $T^{0\nu}_{1/2} > 2.4 \times 10^{25}$ yr. A second, Bayesian analysis was performed, using a flat prior on the inverse half-life in the $0 - 10^{-24}\,\text{yr}^{-1}$ range. The best fit was again zero counts corresponding to a lower limit of $T^{0\nu}_{1/2} > 1.9 \times 10^{25}$ yr at 90% credible interval. The median sensitivity is $T^{0\nu}_{1/2} > 2.0 \times 10^{25}$ yr. The profile likelihood fit was also extended to include the energy spectra from IGEX and HdM experiments, giving a lower limit of $T^{0\nu}_{1/2} > 3.0 \times 10^{25}$ yr at 90% confidence level. Constant background for all five datasets and gaussian peaks with a common half-life were assumed.

In order to compare the GERDA result with the signal claim, a hypothesis test was performed. The expected number of counts for the background only hypothesis, H_0, is 2 ± 0.3 in the $\pm\sigma$ window around $Q_{\beta\beta}$. As an alternative hypothesis, H_1, the claimed signal corresponding to a half-life of $T^{0\nu}_{1/2} = 1.19 \times 10^{25}$ yr plus a background was considered, corresponding to 5.9 ± 1.4 expected counts. In figure 10, the exposure corrected expectation according to the signal claim is shown in dotted red line together with the lower limit derived from the GERDA analysis in blue. The number of observed counts is 3. Assuming the model H_1, the probability to obtain zero counts as the best fit from the profile likelihood analysis is 0.01. Also the Bayes factor, i.e. the ratio of the probabilities of the two models $P(H_1)/P(H_0)$, computed with the GERDA result alone as well as with the combined result is 0.024 and 2×10^{-4}, respectively, therefore the claim is strongly disfavoured. This comparison is restricted to the result of [9] and not [18], due to inconsistencies in the latter, pointed out in [19].

A comparison to the recent limits on the half-life of ^{136}Xe from KamLAND-Zen [20] and EXO-200 [21] is possible, assuming that the leading mechanism is the exchange of a light Majorana neutrino. The experimental results, the claimed signal and the different NME calculations are shown in figure 11.

7. GERDA Phase II

GERDA Phase II aims to improve the half-life sensitivity by another order of magnitude. The sensitivity as a function of the exposure for different background levels is shown in figure 12. An

Journal of Physics: Conference Series **606** (2015) 012006 doi:10.1088/1742-6596/606/1/012006

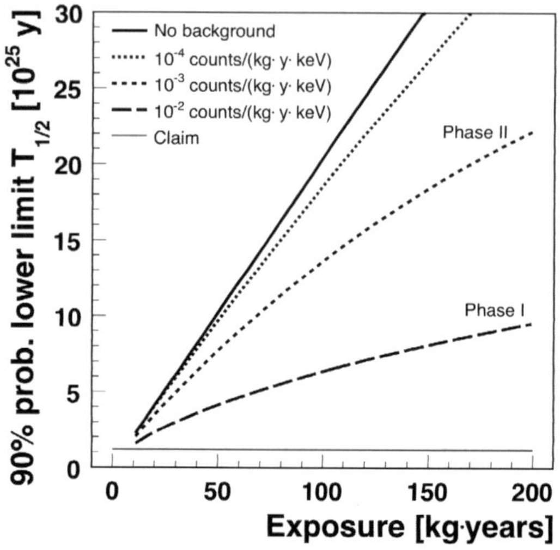

Figure 12. Sensitivity as a function of the exposure for different background cases. The lines corresponding to the background index of Phase I and Phase II are indicated on the figure. Taken from [22].

order of magnitude improvement on the $0\nu\beta\beta$ half-life sensitivity is expected in approximately 5 years.

The size of the detector array is increased to 7 strings. The detectors are assembled in a dry nitrogen atmosphere. The new cable chain is made of selected stainless steel of low radioactivity. The Phase II cables exhibit more than a factor of 10 lower ^{228}Th and ^{226}Ra activities compared to Phase I cables.

The liquid argon will be instrumented with a scintillation background veto system. PMT arrays are installed above and below the detector array. Silicon photomultipliers coupled to wavelength shifting fibres surround the detector array. They will provide increased background reduction capability by detecting scintillation light in liquid argon. Pulse shape analysis in combination with the liquid argon veto provide a suppression factor of 5.2×10^3 at $Q_{\beta\beta}$ for a close ^{228}Th source.

For Phase II, 30 new BEGe detectors were produced. A significant amount of copper and PTFE, for the detector modules, has been replaced by intrinsically radio pure silicon. The energy resolution (FWHM) of the detectors was determined with a ^{60}Co source to be less than 1.9 keV at 1.3 MeV in vacuum. In addition, the A/E pulse shape discrimination, described in section 5, is a robust, simple and well-understood method of background rejection that was successfully implemented during Phase I. Finally, careful handling of the detectors during manufacturing and transportation insures a very low background contribution from ^{60}Co and ^{68}Ge due to cosmogenic activation.

8. Conclusions

Phase I of the GERDA experiment was completed successfully and the design goals were reached. A total exposure of 21.6 kg yr was accumulated. The background index at $Q_{\beta\beta}$ after pulse shape analysis was 0.01 cts/keV kg yr. A blinded analysis looking for the $0\nu\beta\beta$ decay of ^{76}Ge was performed. No signal was observed and the most competitive limit on the half-life of this process for ^{76}Ge was derived, strongly disfavouring the long standing claim of $0\nu\beta\beta$ signal observation.

The transition to GERDA Phase II is ongoing. An additional 20 kg of detector mass will be deployed. The new custom-made BEGe detectors have an excellent pulse shape discrimination capability and a subset of them was tested successfully during Phase I. A liquid argon instrumentation surrounding the detector array will be utilised for further background reduction. The background target of GERDA Phase II is 10^{-3} cts/keV kg yr, which will allow the exploration of ^{76}Ge $0\nu\beta\beta$ half-life values in the 10^{26} yr range.

References

[1] Barabash A S 2010 *Phys. Rev. C* **81**(3) 035501 URL http://link.aps.org/doi/10.1103/PhysRevC.81.035501

[2] Tretyak V I and Zdesenko Y G 2002 *Atomic Data and Nuclear Data Tables* **80** 83 – 116 ISSN 0092-640X URL http://www.sciencedirect.com/science/article/pii/S0092640X01908733

[3] Bilenky S M and Giunti C 2012 *Modern Physics Letters A* **27** 1230015 (*Preprint* http://www.worldscientific.com/doi/pdf/10.1142/S0217732312300157) URL http://www.worldscientific.com/doi/abs/10.1142/S0217732312300157

[4] Vergados J D, Ejiri H and Šimkovic F 2012 *Reports on Progress in Physics* **75** 106301 URL http://stacks.iop.org/0034-4885/75/i=10/a=106301

[5] Rodejohann W 2012 *Journal of Physics G: Nuclear and Particle Physics* **39** 124008 URL http://stacks.iop.org/0954-3899/39/i=12/a=124008

[6] Klapdor-Kleingrothaus H *et al.* (Heidelberg-Moscow Collaboration) 2001 *The European Physical Journal A - Hadrons and Nuclei* **12** 147–154 ISSN 1434-6001 URL http://dx.doi.org/10.1007/s100500170022

[7] Aalseth C E *et al.* (IGEX Collaboration) 2002 *Phys. Rev. D* **65**(9) 092007 URL http://link.aps.org/doi/10.1103/PhysRevD.65.092007

[8] Aalseth C E *et al.* (IGEX Collaboration) 2004 *Phys. Rev. D* **70**(7) 078302 URL http://link.aps.org/doi/10.1103/PhysRevD.70.078302

[9] Klapdor-Kleingrothaus H, Krivosheina I, Dietz A and Chkvorets O 2004 *Physics Letters B* **586** 198 – 212 ISSN 0370-2693 URL http://www.sciencedirect.com/science/article/pii/S0370269304003235

[10] Ackermann K H *et al.* (GERDA Collaboration) 2013 *The European Physical Journal C* **73** 2330 ISSN 1434-6044 URL http://dx.doi.org/10.1140/epjc/s10052-013-2330-0

[11] Pandola L, Bauer M, Krninger K, Liu X, Tomei C, Belogurov S, Franco D, Klimenko A and Knapp M 2007 *Nuclear Instruments and Methods in Physics Research Section A: Accelerators, Spectrometers, Detectors and Associated Equipment* **570** 149 – 158 ISSN 0168-9002 URL http://www.sciencedirect.com/science/article/pii/S0168900206018031

[12] Agostini M *et al.* (GERDA Collaboration) 2014 *The European Physical Journal C* **74** 2764 ISSN 1434-6044 URL http://dx.doi.org/10.1140/epjc/s10052-014-2764-z

[13] Agostini, M *et al.* 2015 *Eur. Phys. J. C* **75** 39 URL http://dx.doi.org/10.1140/epjc/s10052-014-3253-0

[14] Palioselitis D 2014 *Proceedings of the 2nd Workshop on Germanium-Based Detectors and Technologies (Experience from operating germanium detectors in GERDA)*

[15] Agostini M *et al.* (GERDA collaboration) 2013 *Journal of Physics G: Nuclear and Particle Physics* **40** 035110 URL http://stacks.iop.org/0954-3899/40/i=3/a=035110

[16] Agostini M *et al.* (GERDA Collaboration) 2013 *The European Physical Journal C* **73** 2583 ISSN 1434-6044 URL http://dx.doi.org/10.1140/epjc/s10052-013-2583-7

[17] Agostini M *et al.* (GERDA Collaboration) 2013 *Phys. Rev. Lett.* **111**(12) 122503 URL http://link.aps.org/doi/10.1103/PhysRevLett.111.122503

[18] Klapdor-Kleingrothaus H V and Krivosheina I V 2006 *Modern Physics Letters A* **21** 1547–1566 (*Preprint* http://www.worldscientific.com/doi/pdf/10.1142/S0217732306020937) URL http://www.worldscientific.com/doi/abs/10.1142/S0217732306020937

[19] Schwingenheuer B 2013 *Annalen der Physik* **525** 269–280 ISSN 1521-3889 URL http://dx.doi.org/10.1002/andp.201200222

[20] Gando A *et al.* (KamLAND-Zen Collaboration) 2013 *Phys. Rev. Lett.* **110**(6) 062502 URL http://link.aps.org/doi/10.1103/PhysRevLett.110.062502

[21] Auger M *et al.* (EXO Collaboration) 2012 *Phys. Rev. Lett.* **109**(3) 032505 URL http://link.aps.org/doi/10.1103/PhysRevLett.109.032505

[22] Caldwell A and Kröninger K 2006 *Phys. Rev. D* **74**(9) 092003 URL http://link.aps.org/doi/10.1103/PhysRevD.74.092003

2nd Workshop on Germanium Detectors and Technologies | IOP Publishing
Journal of Physics: Conference Series **606** (2015) 012007 | doi:10.1088/1742-6596/606/1/012007

Experience from operating germanium detectors in GERDA.

Dimitrios Palioselitis
on behalf of the GERDA collaboration
Max-Planck-Institut für Physik, München
E-mail: dimitris@mpp.mpg.de

Abstract. Phase I of the Germanium Detector Array (GERDA) experiment, searching for the neutrinoless double beta ($0\nu\beta\beta$) decay of ^{76}Ge, was completed in September 2013. The most competitive half-life lower limit for the $0\nu\beta\beta$ decay of ^{76}Ge was set ($T^{0\nu}_{1/2} > 2.1 \cdot 10^{25}$ yr at 90% C.L.). GERDA operates bare Ge diodes immersed in liquid argon. During Phase I, mainly refurbished semi-coaxial high purity Ge detectors from previous experiments were used. The experience gained with handling and operating bare Ge diodes in liquid argon, as well as the stability and performance of the detectors during GERDA Phase I are presented. Thirty additional new enriched BEGe-type detectors were produced and will be used in Phase II. A subgroup of these detectors has already been used successfully in GERDA Phase I. The present paper gives an overview of the production chain of the new germanium detectors, the steps taken to minimise the exposure to cosmic radiation during manufacturing, and the first results of characterisation measurements in vacuum cryostats.

1. Introduction

GERDA [1] is an experiment looking for the neutrinoless double beta decay of ^{76}Ge. The observation of $0\nu\beta\beta$ decay would imply lepton number violation by two units, indicating physics beyond the Standard Model of particle physics. Additionally, it would prove that neutrinos have a Majorana mass component. Assuming the exchange of light Majorana neutrinos an effective neutrino mass can be evaluated by using predictions for the nuclear matrix elements. The experimental signature of $0\nu\beta\beta$ decay is a peak at the Q-value of the decay at 2039 keV, which lies above the continuous energy spectrum of the neutrino accompanied double beta ($2\nu\beta\beta$) decay. GERDA operates bare germanium diodes inside liquid argon which serves as a coolant and as shielding. The Phase I of GERDA was completed successfully, setting the most competitive limit on the half-life of $0\nu\beta\beta$ decay for ^{76}Ge, $T^{0\nu}_{1/2} > 2.1 \cdot 10^{25}$ yr at 90% C.L. [2, 3].

The efficiency to detect the $0\nu\beta\beta$ decay is increased by the fact that germanium serves both as the source of the decay and as the detector. Additionally, operating the diodes bare in liquid argon provides reduced background from cladding material. Finally, germanium is intrinsically pure and germanium diodes provide excellent energy resolution, while their production chain is well-established. This paper describes the experience obtained by operating germanium diodes in GERDA Phase I as well as the first characterisation results for a subset of the Phase II detectors.

In section 2, an overview of the semi-coaxial detectors used in Phase I and their performance is presented. The description of the broad energy germanium (BEGe) detectors that will be

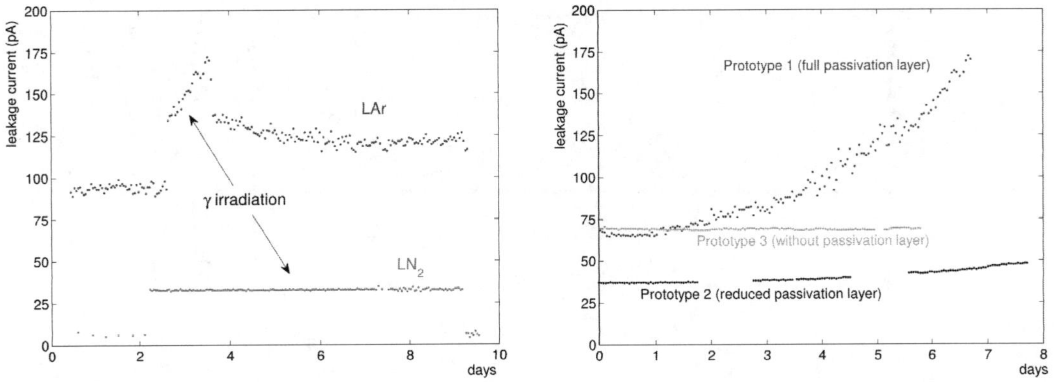

Figure 1. *Left:* Leakage current induced by γ-radiation in a detector operated in liquid argon and liquid nitrogen. *Right:* Leakage current induced by γ-radiation in 3 detectors with different passivation layers. Taken from [1].

used in Phase II, their production chain as well as their performance and characterisation in vacuum and in liquid argon are discussed in section 3.

2. Overview of Phase I semi-coaxial detectors

The main detectors used during GERDA Phase I were 8 enriched semi-coaxial p-type high purity germanium detectors, taken from the previous HdM [4] and IGEX [5] experiments. They are enriched to $\sim$ 86% in ^{76}Ge and have a total mass of 17 kg. The non-enriched detectors have an abundance of $\sim$ 7.8% in ^{76}Ge [1].

The BEGe detectors have a wrap around n+ electrode and a boron implanted p+ contact, separated by an insulated groove (see figure 2). No passivation layer was used on the groove. The masses of the individual detectors range between 0.98 and 2.9 kg. Apart from these 8 coaxial detectors, 5 BEGe detectors [6] were used and tested during Phase I. The energy resolution of the semi-coaxial detectors at 2.6MeV, determined with a ^{228}Th calibration source, is 4-5 keV (FWHM) [1]. Two detectors developed high leakage current at the beginning of Phase I and were removed from the analysis, however, they were used as a veto to suppress multi-site events.

A long term study of the leakage current of bare natural germanium detectors operated in LAr and LN_2 under varying gamma-irradiation conditions was performed. A continuous increase of the leakage current was observed during irradiation for the detector in LAr but not in LN_2, as shown on the left panel in figure 1. After removing the source, the leakage current would stabilise again, but at a value higher than before the irradiation. The leakage current could be completely restored to its initial value by warming up the detector in methanol baths. The right plot of figure 1 shows measurements with three prototype detectors using different sizes of groove passivation, namely large area, reduced and none. Reducing the size of passivation layer suppresses the effect while there is no leakage current induced when there is no passivation layer.

It is believed that the leakage current increase is induced by the collection and trapping of charges produced by the ionisation of LAr on the passivated surface of the detectors. Therefore, all GERDA Phase I detectors were reprocessed without the evaporation of a passivation layer [1].

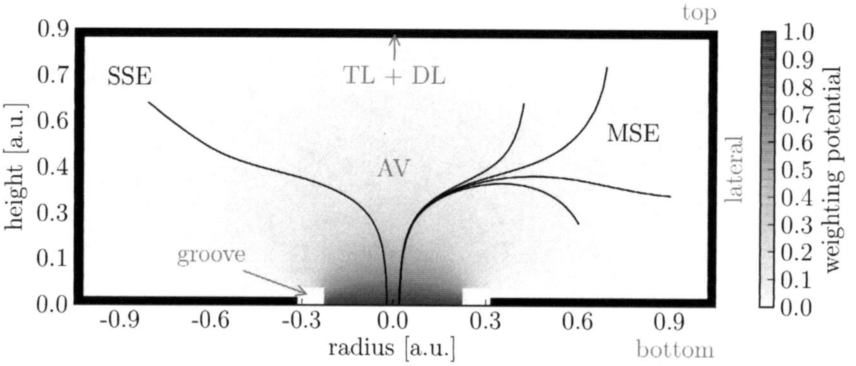

Figure 2. Profile and weighting potential of a BEGe diode. The weighting potential is strongest close to the p+ read-out electrode. The p+ electrode (centre,bottom) is separated from the n+ electrode by a groove, covered with an insulating passivation layer. The inactive dead layer (DL), the transition layer (TL) and the active volume (AV) are indicated. Examples of SSE and MSE events by means of hole trajectories are also shown. Taken from [7].

3. Phase II BEGe detectors

BEGe detectors consist the majority of the detectors to be used during Phase II of GERDA. These detectors are made from p-type germanium enriched to 88% in ^{76}Ge, they have a wrap around n+ electrode, and a p+ electrode acting as an electron blocking contact. A reduced passivation layer was used for the BEGE detectors. Five out of the 30 new detectors were deployed and tested during Phase I. Compared to the Phase I coaxial detectors, the BEGe detectors have a smaller size, lower noise and better energy resolution as well as enhanced pulse shape discrimination capability.

A schematic of a BEGe diode, along with the weighting potential profile is shown in figure 2. The potential peaks strongly close to the p+ electrode. The hole trajectories of individual energy depositions for a single-site event (SSE) and a multi-site event (MSE) are illustrated.

In Phase II, electrical contacts to the point contact are made using ultra sonic wire bonds. This has the advantage of having low mass electrical contact, allowing the reduction of the mass of the detector holders, consequently lowering the background contribution from excess cladding material. For the bonding, a 600 nm thin Al film is deposited on the diodes to provide a metallic surface for ohmic contact and good bonding. This contacting scheme requires stability of the wires in LAr, as well as their survival during warming and cooling cycles. Special care during bonding is necessary to prevent damaging the p+ contact.

3.1. BEGe detector production

The enrichment in ^{76}Ge was performed in 2005 at ECP in Zelengorsk, Russia [8], where 53.3 kg of enriched germanium oxide powder were produced. Improvements of clean conditions at work places and the use of de-ionised water for hydrolysis led to a purity level of 99.99%. The first step in the enrichment procedure is the fluorination of natural germanium, i.e. binding of Ge in gaseous GeF_4 compounds, followed by enrichment via gas centrifuge process and finally production of the enriched germanium oxide powder after hydrolysis, drying and calcination. Reduction and zone refinement was done in 2010 by PPM in Langelsheim, Germany [9], resulting in 37.5 kg of 6N purity grade germanium. The germanium oxide powder was reduced to metallic germanium in H_2 atmosphere, followed by cleaning, etching and zone-refinement of the metal ingots. The low resistivity tails of the ingots underwent additional zone refinement. Finally,

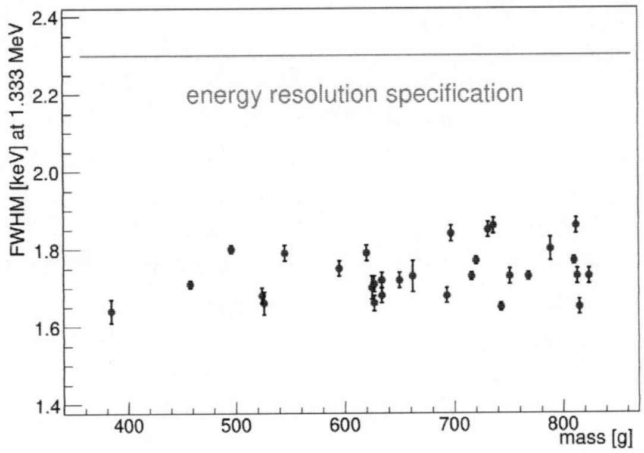

Figure 3. Energy resolution (FWHM) of the 30 Phase II BEGe detectors as a function of their mass determined at the 1333 keV ^{60}Co peak. The error bars indicate the uncertainties from the fit of the peaks. Taken from [7].

crystal pulling and cutting took place in Canberra, Oak Ridge [10], where 9 crystal ingots were cut into 30 slices, with a total mass of 20.8 kg.

The conversion of the germanium crystal slices to detectors was done in 2012 at Canberra Semiconductors N.V., Olen, Belgium [11]. Thirty enriched germanium diodes were produced, corresponding to 20 kg. This is a mass yield of 53.3% from the original germanium oxide. Prior to delivery, Canberra tested the diodes in liquid nitrogen bath for basic parameters. The requirements set on energy resolution (< 2.3 keV), operational voltage (< 3 kV) and leakage current (< 50 pA) were met for 29 out of 30 detectors. The energy resolution for all 30 detectors as a function of their mass is shown in figure 3. The mean energy resolution at 1333 keV is 1.74 keV (FWHM). One of the detectors could not reach full and stable depletion voltage and therefore had a deteriorated charge collection efficiency in some parts of the crystal.

3.2. Cosmic activation

An important background for $0\nu\beta\beta$ searches originates from cosmic activation of germanium during the production and handling of the germanium diodes. At sea level, secondary cosmic rays consisting mainly of fast nucleons, muons and muon induced neutrons produce long lived radionuclides via spallation reactions. The subsequent decays of these isotopes are a source of background in a germanium $0\nu\beta\beta$ experiment. The relevant isotopes produced in germanium are ^{68}Ge and ^{60}Co, with half-lives of 270 days and 5.2 years respectively.

The contribution of these decays to the background index (B.I.) of Phase II was studied with Monte Carlo simulations [12, 13]. Averaged over three years of data taking, ^{68}Ge and ^{60}Co contribute approximately 3.7×10^{-3} counts/keV kg yr and 8.4×10^{-4} counts/keV kg yr. This level of B.I. can be reduced by one to two orders of magnitude after applying pulse shape discrimination methods. The goal of GERDA Phase II is to reduce the background index by an order of magnitude compared to Phase I, reaching the level of 10^{-3}cts/keV kg yr.

Several actions were taken to minimise the exposure to cosmic radiation during the production of the diodes. Optimisation of the processing steps minimising the time the material was unshielded, on-site underground storage during the various processing steps and using a water- and steel-shielded container during transportation contribute to the reduction of cosmogenic-induced radioisotopes in the detectors. The expected reduction factor of cosmic activation due to the shielding was in the range of 10 and 15 for ^{68}Ge and ^{60}Co, respectively [14]. Figure 4 shows the history of the estimated number of cosmogenic-induced ^{68}Ge and ^{60}Co atoms in enriched germanium for a BEGe detector. The bulk of the activation occurred during enrichment (I), purification (II), crystal growth (III) and diode conversion (IV), while the material was above ground. ^{60}Co was removed completely at the time of crystal growth.

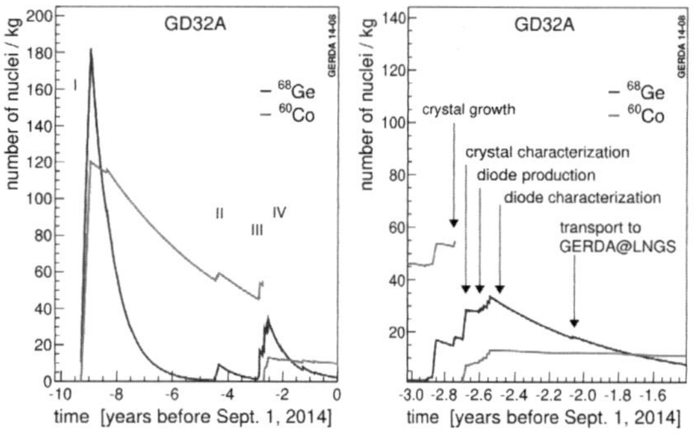

Figure 4. *Left:* Estimated number of cosmogenic-induced ^{60}Co and ^{68}Ge atoms in ^{enr}Ge as a function of time, during the production of one of the BEGe diodes. Increased activation takes place during (I) enrichment, (II) purification, (III) crystal growth and (IV) diode conversion. *Right:* Zoom of the activation history of the detector during crystal growth and diode conversion. Taken from [7].

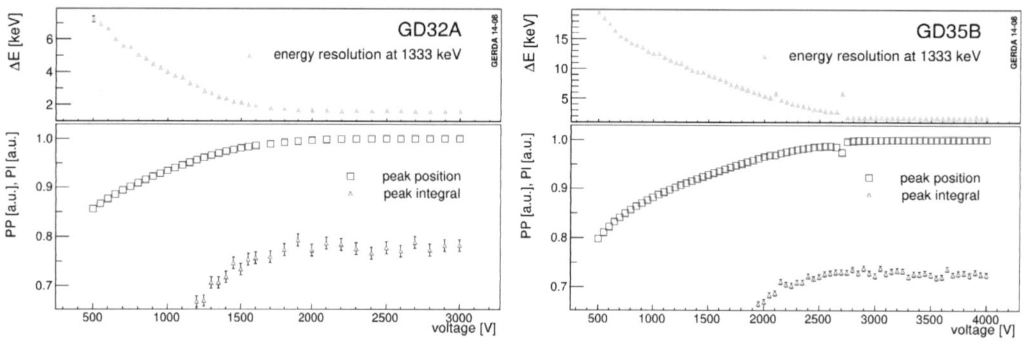

Figure 5. ^{60}Co high voltage (HV) scans for detectors GD32A (left) and GD35B (right). Energy resolution ΔE (FWHM at 1333 keV), peak position PP and peak integral PI as a function of the applied voltage. Taken from [7].

3.3. Vacuum cryostat tests

The following characterisation measurements took place in the HADES underground laboratory in Mol, Belgium. High voltage scans, in steps of (50-100) V, using a point like ^{60}Co source were performed to the first seven BEGe detectors in order to determine the depletion voltage. The depletion voltage of a germanium detector is defined as the reverse bias voltage that electrically fully depletes the diode of free charge carriers. The peak position, the peak integral and the energy resolution in terms of FWHM of the two ^{60}Co γ-lines were monitored at each intermediate voltage.

The HV scans for two detectors are shown in figure 5. As soon as the detector is electrically full depleted, the parameters become almost constant. The lowest voltage point on this plateau is the depletion voltage.

The measurements in HADES show energy resolutions which are in good agreement with the values provided by Canberra [7]. Additionally, the BEGe detectors show a 30% better energy resolution compared to the coaxial detectors operated in vacuum cryostats [15]. In particular, the average energy resolution (FWHM) at the recommended voltage for the 1333 keV and 2615 keV gamma lines are 1.73 keV and 2.47 keV respectively.

In figure 5 one sees that the detectors exhibit the so-called bubble or pinch-off effect [6, 16]. In a voltage interval of a few tens of kV below depletion voltage, the electric field in a region at the centre of the detector becomes zero. Some clouds drifting to the electrodes may get trapped in this region, giving rise to a broader energy resolution and a peak position instability right below the depletion voltage.

The active volumes of the seven BEGe detectors were determined by comparing calibration

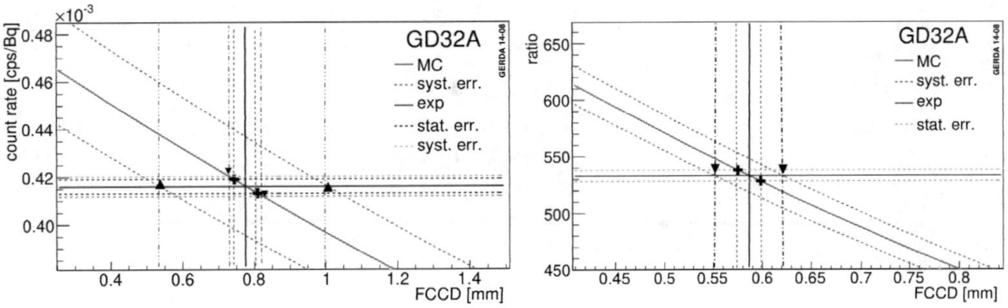

Figure 6. Evaluation of the FCCD for detector GD32A. *Left:* Measured vs. simulated absolute count rate of the 1173 keV ^{60}Co γ-line. *Right:* Measured vs. simulated ratio of two γ-line intensities for ^{241}Am. Taken from [7].

data to simulated calibrations of the same experimental setup. Two different methods were used. In one case a ^{60}Co source was used and the count rate for a subset of gamma-ray peaks was measured. In the second method, an ^{241}Am source was used and count rate ratios were determined. Both observables are plotted as a function of the full charge collection depth (FCCD), i.e. the sum of transition layer and dead layer thicknesses [7], based on simulations (see figure 6). The transition layer surrounds the active volume (AV), has a reduced charge collection efficiency and a low electric field [17]. The intersection of the experimental data with the curves from the simulations provide the FCCD of the detector. The active volume fractions are determined by subtracting the FCCD volume from the overall detector volume. It is worth noting that the peak count rate method results depend on the precise knowledge of the activity of the source while this factor cancels out in the peak ratio method.

The active volume fractions of the detectors lie on the range of 89-94% of the detector volume [7]. The results from the ^{241}Am measurements are in good agreement, within 1%, with the values provided by the manufacturer, determined also with Am surface scans. The results from ^{60}Co measurements are systematically lower than the ^{241}Am results by 1.9% on average. Details on the systematic uncertainties of these measurements are reported in [7].

For the determination of background rejection efficiencies through pulse shape analysis, ^{228}Th calibration sources were deployed on the outer surface of the vacuum cryostat. As an SSE proxy, the 1593 keV double escape peak (DEP) from the 2614 keV line of ^{208}Tl was used. Events from the full energy peak (FEP), the single escape peak (SEP) and Compton continua are taken as MSE proxies. The pulse shape discrimination (PSD) method used for the BEGe detectors utilises a single parameter, i.e. the maximum amplitude of the charge pulse, A, over the energy deposited, E [18].

Figure 7 shows the A/E distributions of DEP events from ^{208}Tl decays for three BEGe detectors. Ideally, as shown in black, the distribution is Gaussian with a tail component from underlying MSE background. Five BEGe detectors showed anomalous A/E distributions, with multiple or unusually broad peaks. The A/E distributions for two of these detectors are shown in red and blue in figure 7. Investigations indicated that this behaviour is not caused by the set-up configuration (e.g. electronic components) [7]. However, positively charged compounds can be deposited in the groove between the p+ and n+ electrodes after diode production leading to the distortion of the electric field configurations and in turn distorted drift paths. The grooves of two detectors underwent chemical treatment by the manufacturer followed by thermal heating. This led to improvement of the PSD behaviour for one of the two detectors.

Journal of Physics: Conference Series **606** (2015) 012007 doi:10.1088/1742-6596/606/1/012007

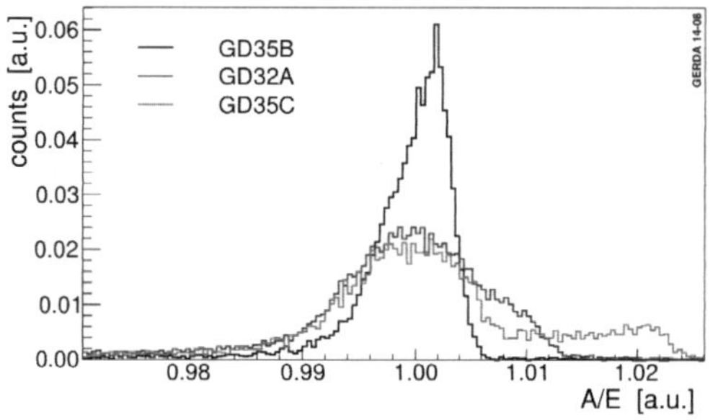

Figure 7. Normalised A/E distributions of DEP events from ^{208}Tl decays for three BEGe detectors. Taken from [7].

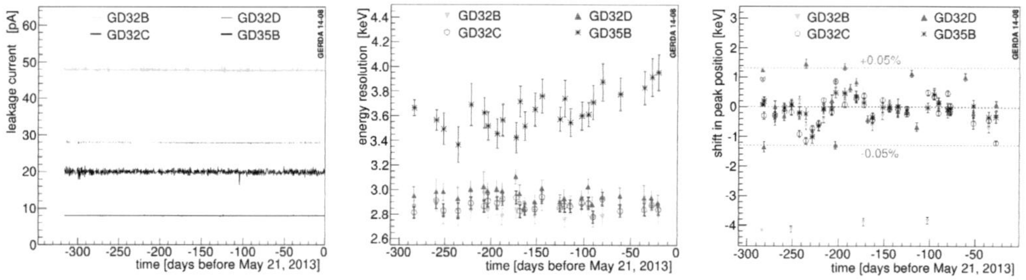

Figure 8. Operational stability of BEGe detectors operated in liquid argon during GERDA Phase I. *Left:* Leakage current stability. *Middle:* Energy resolution (FWHM of the 2615 keV ^{208}Tl γ-line) as a function of time. *Right:* Variability of the 2615 keV γ-line peak position. Taken from [7].

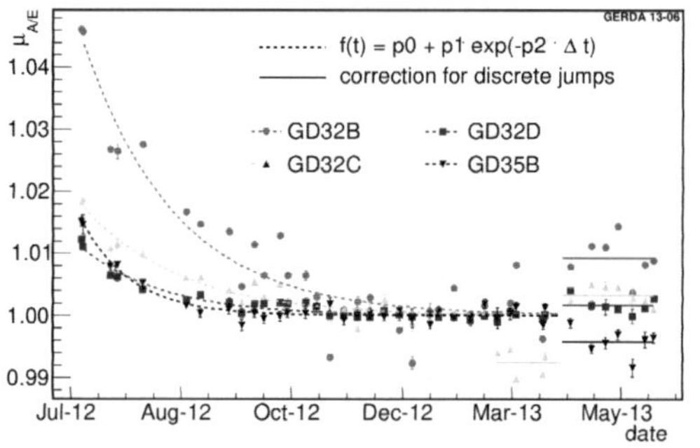

Figure 9. Gaussian mean of the A/E distributions of DEP events from individual ^{228}Th calibrations as a function of time. An exponential fit was performed to correct the A/E drift for data and calibration. Jumps that occurred after a power failure were also corrected. Taken from [18]

3.4. Performance in liquid argon

The performance of four BEGe detectors operated in LAr during Phase I of GERDA is shown in figure 8. The leakage current, shown on the left plot, is stable. The energy resolution, determined with ^{228}Th calibration sources, is also stable, with values in the range between 2.8 to 3 keV. This resolution is 30% worse than what was determined with the detectors operating in vacuum. This is due to the longer signal cable between the readout electrode and the field effect transistor (FET) in liquid argon. Compared to the coaxial detectors, the resolution of the BEGes is 30% better [12]. Finally, the peak position of the 2615 ^{208}Tl line is stable over time, with relative deviations of 0.1%.

During Phase I, a drift of the mean of the A/E distributions was observed. First, a long term exponentially decreasing behaviour, shown in figure 9, varying from 1 to 5% depending on the detector. Second, a shift of 1% to higher values was observed during ^{228}Th calibrations. The instabilities were quantified and time-corrections were applied to the A/E distributions. The

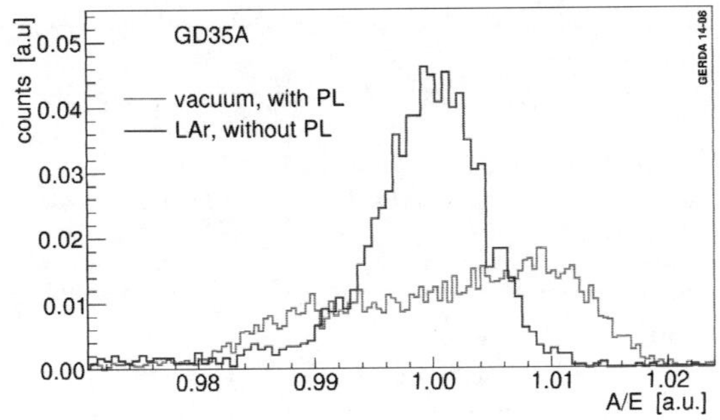

Figure 10. Normalised A/E distributions of DEP events measured with detector GD35A with passivation layer in vacuum (red) and without passivation layer in liquid argon (blue). Taken from [7].

A/E drift might be due to the presence of charges on the passivated groove, neutralising or dissolving in liquid argon after months of operation.

Finally, in order to examine the pulse shape anomalies observed in vacuum, tests were performed in liquid argon in the GERDA Detector Laboratory (GDL) in LNGS. In liquid argon tests, the width of the A/E distributions improved and the non-Gaussian features disappeared [7]. The PSD survival efficiencies were also improved for the detectors without passivation layer and no A/E anomaly present. Figure 10 shows the A/E distributions for a detector with passivation layer in vacuum and without passivation layer in LAr.

4. Conclusions

The goals of GERDA Phase I were reached with well-type non-passivated coaxial detectors, setting a limit on the half-life of ^{76}Ge $0\nu\beta\beta$ decay, as well as demonstrating the successful operation of bare germanium detectors in LAr. Irradiation induced leakage current was investigated and was found to be strongly suppressed by reducing the size of the passivation layer. For Phase II, 30 new BEGe detectors were produced. Careful steps were taken during production to minimise cosmic activation, that can lead to background at the region of interest. A new contacting scheme is being implemented for the BEGe detectors in Phase II. The new detectors have an improved energy resolution and high background identification capabilities. A possible origin of the anomalies related to pulse shape analysis is charge deposition in the groove. Improvement can be achieved by chemical and thermal treatment of the passivation layer or by removing the passivation and handling the detectors under clean nitrogen atmosphere. The BEGe detectors were proven to be very stable in their operation in liquid argon. The low expected background index in Phase II will allow the exploration of $0\nu\beta\beta$ half-life values of ^{76}Ge above 10^{26} years.

References

[1] Ackermann K H *et al.* (GERDA Collaboration) 2013 *The European Physical Journal C* **73** 2330 ISSN 1434-6044 URL `http://dx.doi.org/10.1140/epjc/s10052-013-2330-0`

[2] Agostini M *et al.* (GERDA Collaboration) 2013 *Phys. Rev. Lett.* **111**(12) 122503 URL `http://link.aps.org/doi/10.1103/PhysRevLett.111.122503`

[3] Palioselitis D 2014 *Proceedings of the 2nd Workshop on Germanium-Based Detectors and Technologies (Results on neutrinoless double beta decay of ^{76}Ge from GERDA Phase I)*

[4] Klapdor-Kleingrothaus H *et al.* (Heidelberg-Moscow Collaboration) 2001 *The European Physical Journal A - Hadrons and Nuclei* **12** 147–154 ISSN 1434-6001 URL `http://dx.doi.org/10.1007/s100500170022`

[5] Aalseth C E *et al.* (IGEX Collaboration) 2002 *Phys. Rev. D* **65**(9) 092007 URL `http://link.aps.org/doi/10.1103/PhysRevD.65.092007`

[6] Budjáš D *et al.* 2009 *Journal of Instrumentation* **4** P10007 URL `http://stacks.iop.org/1748-0221/4/i=10/a=P10007`

[7] Agostini, M *et al.* 2015 *Eur. Phys. J. C* **75** 39 URL `http://dx.doi.org/10.1140/epjc/s10052-014-3253-0`

[8] Joint Stock Company "Production Association Electrochemical Plant", Pervaya Promyshlennaya 1, 663690 Zelenogorsk, Russia `http://www.ecp.ru/index en.shtml`
[9] PPM Pure Metals GmbH, Am Bahnhof 1, 38685 Langelsheim, Germany `http://www.ppmpuremetals.de/`
[10] Canberra Industries Inc., 107 Union Valley Rd, Oak Ridge, TN, USA `http://www.canberra.com/`
[11] Canberra Semiconductor N.V., Lammerdries 25, 2250 Olen, Belgium `http://www.canberra.com/`
[12] Agostini M *et al.* (GERDA Collaboration) 2014 *The European Physical Journal C* **74** 2764 ISSN 1434-6044 URL `http://dx.doi.org/10.1140/epjc/s10052-014-2764-z`
[13] Agostini M 2013 *Signal and background studies for the search of neutrinoless double beta decay in GERDA* Ph.D. thesis Technische Universität München URL `http://www.mpi-hd.mpg.de/gerda/public/pub-phd.html`
[14] Barabanov I, Belogurov S, Bezrukov L, Denisov A, Kornoukhov V and Sobolevsky N 2006 *Nuclear Instruments and Methods in Physics Research Section B: Beam Interactions with Materials and Atoms* **251** 115 – 120 ISSN 0168-583X URL `http://www.sciencedirect.com/science/article/pii/S0168583X06006793`
[15] Barnabé M 2009 *Performance and stability tests of bare high purity germanium detectors in liquid argon for the GERDA experiment* Ph.D. thesis University of Heidelberg URL `http://www.mpi-hd.mpg.de/gerda/public/pub-phd.html`
[16] Abgrall N *et al.* (Majorana Collaboration) 2014 *Adv.High Energy Phys.* **2014** 365432 (*Preprint* `1308.1633`)
[17] Aguayo E *et al.* 2013 *Nuclear Instruments and Methods in Physics Research Section A: Accelerators, Spectrometers, Detectors and Associated Equipment* **701** 176 – 185 ISSN 0168-9002 URL `http://www.sciencedirect.com/science/article/pii/S0168900212012752`
[18] Agostini M *et al.* (GERDA Collaboration) 2013 *The European Physical Journal C* **73** 2583 ISSN 1434-6044 URL `http://dx.doi.org/10.1140/epjc/s10052-013-2583-7`

Germanium detector test-stands at the Max Planck Institute for Physics and alpha interactions on passivated surfaces

C Gooch, L Garbini, I Abt, O Schulz, M Palermo, B Majorovits, H-Y Liao, X Liu and H Seitz

Max-Planck-Institute for Physics, Föhringer Ring 6, 80805 München, Germany

gooch@mpp.mpg.de

Abstract. The GeDet group at the Max Planck Institute for Physics in Munich, Germany, operates a number of test stands in order to conduct research on novel germanium detectors. The test stands are of a unique design and construction that provide the ability to probe the properties of new detector types. The GALATEA test stand was especially designed for surface scans, specifically α-induced surface events, a problem faced in low background experiments due to unavoidable surface contamination of detectors. A special 19-fold segmented coaxial prototype detector has already been investigated inside GALATEA with an α-source. A top surface scan provided insight into the physics underneath the passivation layer. Detector segmentation provides a direct path towards background identification and characterisation. With this in mind, a 4-fold segmentation scheme was implemented on a broad-energy point-contact detector and is being investigated inside the groups K1 test stand. A cryogenic test-stand where detectors can be submerged directly in liquid nitrogen or argon is also available. The goal is to establish segmentation as a viable option to reduce background in future large scale experiments.

1. Introduction

The Standard Model of particle physics has withstood the test of time, even though many high precision experiments have tried to find deviations from its predictions over the last two decades. But despite its success, the theory does have a number of shortcomings that require at least some extensions to the current framework. Inconsistency with general relativity, the strong CP problem, the evidence for dark matter (DM) and neutrino oscillations all hint at physics beyond the standard model.

The germanium detector group (GeDet) at the Max Planck Institute for Physics (MPP) develops germanium detectors suitable to search for some of this new physics, so far specifically focusing on neutrinoless double beta decay ($0\nu\beta\beta$). The observation of $0\nu\beta\beta$ would demonstrate Lepton Number Violation and establish the neutrino as a Majorana particle. The isotope ^{76}Ge decays via standard double beta ($2\nu\beta\beta$) decay and is, thus, a candidate for $0\nu\beta\beta$. Detectors made of material enriched in ^{76}Ge are both source and detector.

Germanium detectors are also used for direct DM searches. Weakly interacting massive particles could interact in the germanium crystal, depending on their mass and the cross section. Neither DM or $0\nu\beta\beta$ have so far been directly observed. However, limits have been established in both cases. Although the requirements on the germanium detectors are quite different for the two searches, a goal is to develop a detector that could be used for both searches in a future large scale experiment.

For any such future experiment, the reduction and identification of background events is of utmost importance. Past experience [1] shows that in a well shielded detector the remaining background is caused by radioactivity in materials close to the detectors. The use of well-developed analysis tools and especially designed detectors is essential to suppress this background. Detector segmentation is a special feature used by the GeDet group as a method to identify background and understand detector physics. Detectors are developed in cooperation with Canberra France and are investigated in test-stands which are developed and operated at the MPP in Munich.

2. GALATEA test stand

The principle of the GALATEA [2] test stand is to be able to perform complete scans of the top and side of a germanium detector with no material between source and detector. This allows investigations with minimally penetrating alpha and beta particles. The stages to facilitate the scans are necessarily inside the same vacuum volume as the detector. The preamplifiers are also inside this volume. This in order to keep cable length, and therefore input capacitances to the preamplifiers, to a minimum. A schematic of the GALATEA setup is shown in figure 2.

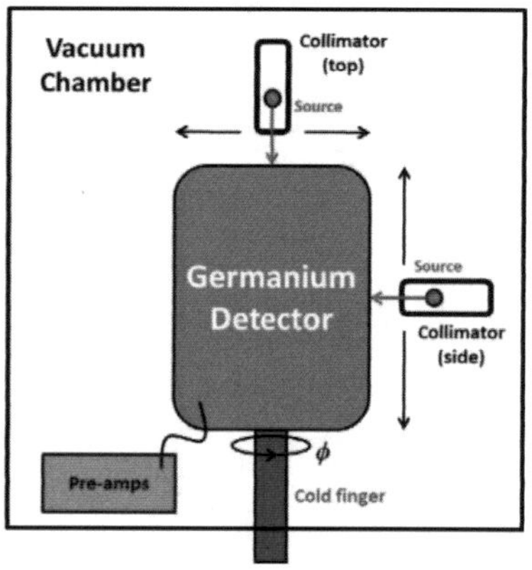

Figure 1. Principle of the GALATEA test stand.

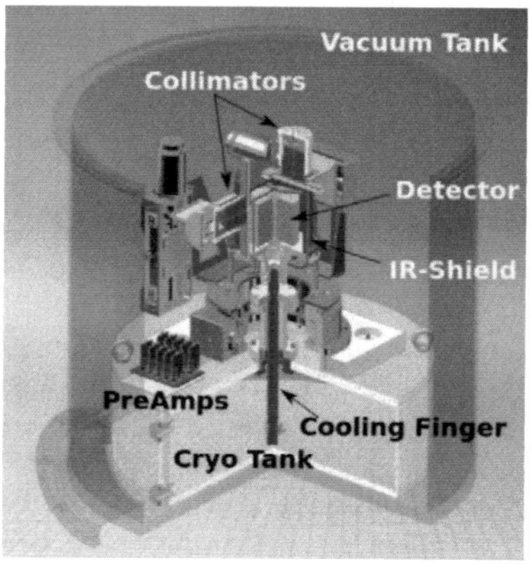

Figure 2. Schematic of the GALATEA setup [2].

Germanium detectors only function at cryogenic temperatures, therefore, the detector holder is connected to a cold finger which itself is connected to a tank filled with liquid nitrogen, also housed within the vacuum chamber. Two linear stages holding tungsten collimators are attached to a rotation stage. One linear stage moves its source radially across the top, the other moves up and down the side. Together with the rotation stage, the complete side and top of the detector can be scanned

3. SuperSiegfried test detector

SuperSiegfried (SuSie) is a cylindrical, true coaxial, n-type high purity germanium (HPGe) detector, which is 18 + 1 fold segmented. This means that the mantle is segmented 3-fold in z and 6-fold in ϕ plus a 19th segment on top which is unsegmented in ϕ, see figure 3. Each segment provides event by event energy (i.e. spectroscopy data) and pulse data. This data can then be crossed check with simulation to help build up a picture of the physics inside the detector. The segments are only partially metallised with each segment read out separately via a kapton cable with snap contacts [3]. The 19th segment is only metallised in a small sector, where the contact is placed.

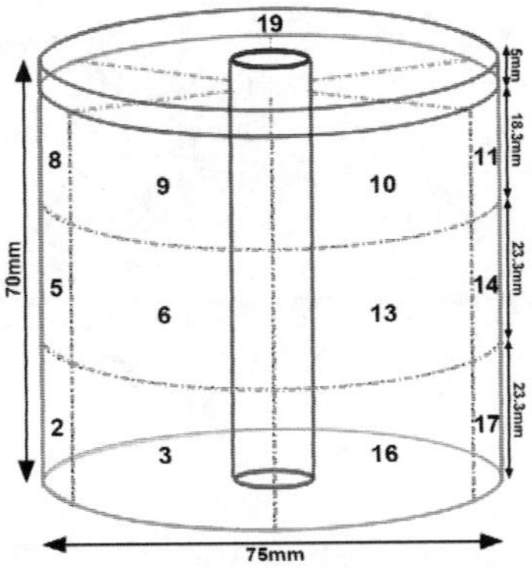

Figure 3. SuSie segmentation scheme [4].

Figure 4. SuSie detector with kapton cable [2] on transport ring.

Detectors with the 18-fold segmentation scheme were used to study the reduction of photon induced background in $0\nu\beta\beta$ experiments. The 5mm thick 19th top segment was designed especially to study surface channel effects and charge trapping underneath the passivation layer on this top plate.

4. Top plate alpha scan

Alpha particles have little penetration power and are therefore used to study events close to the detector surface. An ^{241}Am source producing ≈5.6 MeV alpha particles was used to irradiate the passivated top surface of the SuSie detector. These alpha particles have a nominal penetration depth of ≈30μm in germanium and therefore are contained in the 19th segment. This can be seen clearly in the energy spectrum in figure 5 which shows the spectrum of a measurement at the marked position. The broad bump around 2.5 MeV is due to the alphas and can only be seen in the core (blue) and segment 19 (red) spectrum. The sum of the spectra recorded for segments 1 to 18 (green) show no alpha bump.

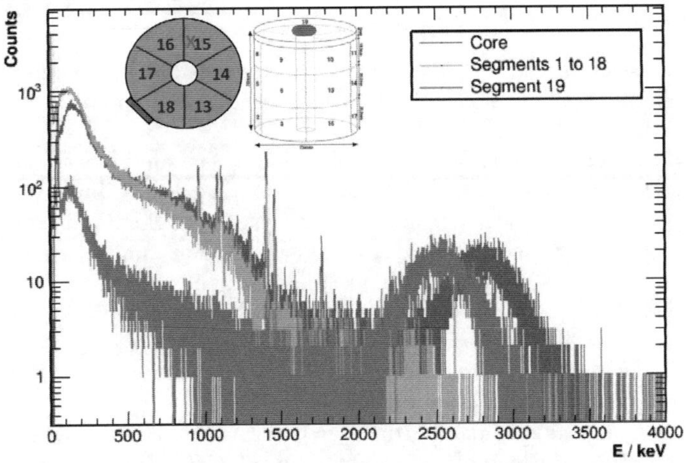

Figure 5. Energy spectra of the core (blue), segment 19 (red) and segments 1 to 18 (green) with a collimated ^{241}Am source aiming down at the top 19th segment at position "X". The difference in the position of the alpha deposition energy peak between the core and 19th segment is clearly visible. The holes and electrons are subject to different trapping effects in this region.

The detector has a dead layer on top consisting of the passivation layer and the area of low field underneath. In principle, only the energy deposited underneath that layer is recorded. In practice things are more complicated. The average energy deposited in the detector i.e. the bump position, is taken as the energy deposited underneath an "effective" dead layer. In the core, a shift of the average energy deposition to ~2.5 MeV from the initial ~5.6 MeV alpha particles is observed. However, figure 5 also shows that the peak position of the bump is different for the 19th segment, demonstrating that the classical picture of well-defined layers is oversimplified. The energy recorded in the core is dominated by electron drift while that of the segments by hole drift. The two different charge carriers are subject to different trapping effects, which is a typical feature of events close to the surface. This effect can also be seen by comparing the recorded pulses from the core and segment 19. Figure 6 shows a smaller pulse in segment 19 than in the core which is indicative of a more incomplete charge collection due to hole trapping than due to electron trapping. The pulse in segment 19 also has a longer rise time in comparison with the core pulse. The reduced speed of the holes could be due to a very low field gradient underneath the top surface at higher radii [5].

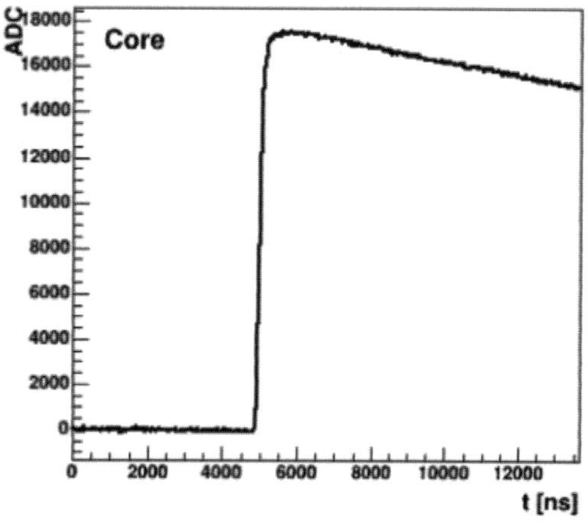

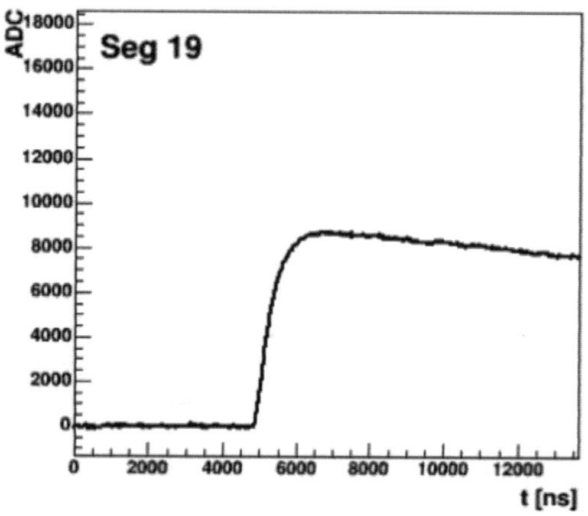

Figure 6. Example pulses from the core and segment 19 from the data sample used for figure 5. The comparatively smaller pulse seen in segment 19 is indicative of hole trapping.

A scan of the top surface was done with the collimated ^{241}Am source positioned at varying radii. The energy spectra of the core and segment 19 readouts for varying radii are shown in figure 7. The average alpha peak energies are clearly dependent on the radial position. The energy seen in the core lessens the further away from the core the source is positioned, meaning that the larger the radii the larger the probability of electrons being trapped. For holes the opposite is true. This radial effect is due to the path length of the charges inside the crystal and the field configuration. At smaller radii the holes have a longer path to the electrode and are more effectively trapped than the electrons and therefore the segment energy is less than the core. However, as the radii increases the electrons become trapped more effectively and the segment energy becomes more than the core. The energy spectra in figure 5 show a situation where the energy recorded in the core is higher than in segment 19 and, therefore, that at this position the holes are being trapped more effectively than the electrons.

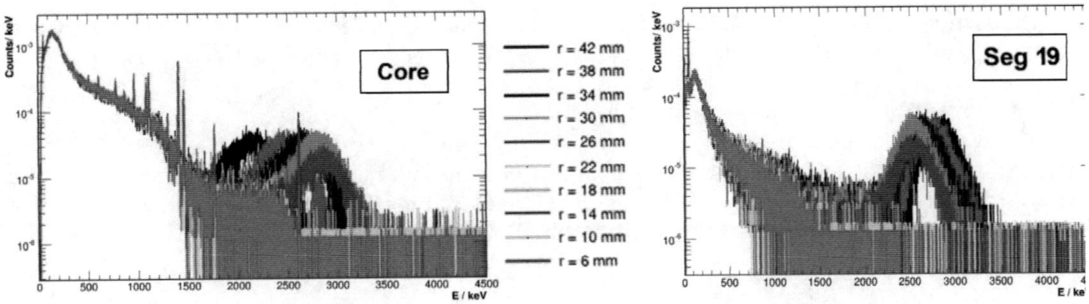

Figure 7. Energy spectra for varying radii with respect to the core position for the core readout (left) and the segment 19 readout (right). The average alpha peak energies are clearly dependent on the radial position.

The different charge trapping effects can be interpreted as effective dead layers for holes and electrons. See figure 8. At small radii, i.e close to the core, no alpha bump can be seen. The effective dead layers are thus more than 30μm, the penetration depth of the alphas, thick. Earlier investigations with photons [5] show dead layers of the order of millimetres. The detailed studies in the region of the tenth of micron layer thicknesses were only possible with alphas as probing particles.

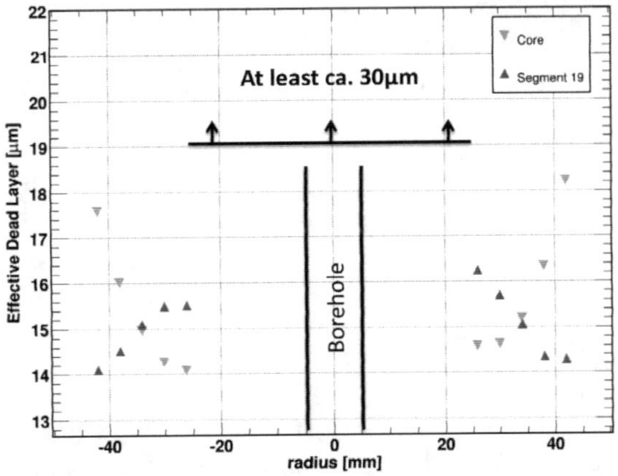

Figure 8. Effective dead-layer thickness for the core (green) and segment 19 (red) as a function of the radial position of the alpha source. The different charge trapping effects result in different effective dead layers for the holes and electrons.

An azimuthal scan was also carried out, i.e. at fixed radius with varying ϕ. The results demonstrated that the metallisation (as marked in figure 5) has a significant effect on the charge collection efficiency. In both the core and segment 19 spectra, the recorded energies shifted to larger energies as the source was moved closer to the metallisation. As the charge collection efficiency increases closer to the metallisation, it suggests increased electric field homogeneity in that region. This was observed earlier [5].

The use of alphas with the SuSie detector is a good way of studying detector surface effects in detail. The difference between the energy seen in the core and the segment provides a background rejection marker without the good bandwith needed for pulse shape analysis. In the future, a complete characterisation of SuSie using alpha, beta and gamma sources will be carried out. GALATEA will also be used to characterise other detector types and further probe germanium detector properties.

5. K1 test stand and a first segmented n-type BeGe detector

The K1 test stand as depicted in figure 9 was built by Canberra France and modified by GeDet in Munich. It is used to study the detector response to photons and neutrons. The set up consists of a conventional cryostat containing one detector cooled via a cooling finger submerged in a LN2 dewar. Two copper "ears" on either side of the cryostat house the preamplifiers.

Figure 9. K1 test stand in Munich.

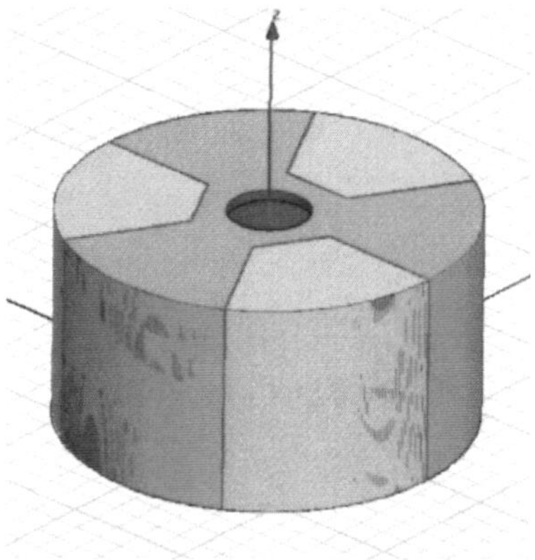

Figure 10. Conceptual design of the segmented BeGe.

In May 2014, the GeDet group received a segmented broad-energy (BeGe) point-contact n-type HPGe detector built by Canberra France. The segmentation scheme, as shown in figure 10, was designed as a compromise between having as few channels and as much capability of position reconstruction as possible. The segmentation is able to break ϕ degeneracy and pulse drift time degeneracy, thus allowing position reconstruction in 3D.

A top scan of the segmented BeGe in the K1 test stand has already been carried out with a ^{133}Ba source. Figure 11 shows averaged pulses for the 81keV line at varying radii of the source position. The averaging helps to demonstrate the features of these low amplitude pulses by eliminating the influence of electronic noise. At 81keV, the minimal variation of the depth of the interaction justifies averaging. The core signals do not change much for varying radii. The signal in the segment collecting the charge, segment 3, gets shorter as the radius decreases. The mantle and the two segments not collecting the charge show clear negative mirror pulses corresponding to the long drift of the electrons to the other end of the cylindrical detector. The mirror pulse in the mantle gets shorter as the radius decreases. However, the amplitude is constant. In the two non-collecting segments, the situation is more involved. At low radii, the amplitude is larger due to the proximity of the segments. However, these mirror pulses are shorter than for larger radii and they start later with respect to the core pulse.

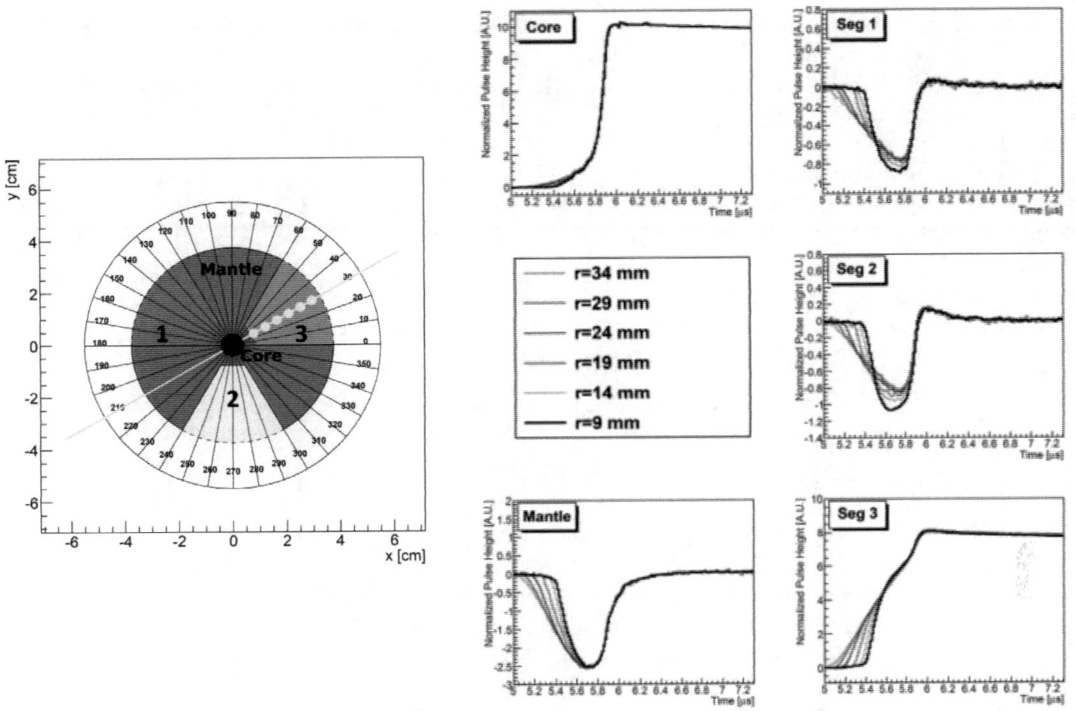

Figure 11. Averaged pulses for the 81keV line of a ^{133}Ba top scan over varying radii of segment 3 of the segmented BeGe. The radial dependence of the pulse length can be seen in the signal in segment 3 and the mirror pulses in segments 1, 2 and mantle. The core signals change very little.

6. Gerdalinchen II

Gerdalinchen II is a cryogenic test facility designed to mimic the conditions in the GERDA experiment. The setup, as shown in figure 12, can hold up to 3 detectors in their holders in a string and submerge them directly into cryogenic liquid. The original goal was to verify that large segmented n-type crystals can be operated in cryogenic liquid over long periods of time.

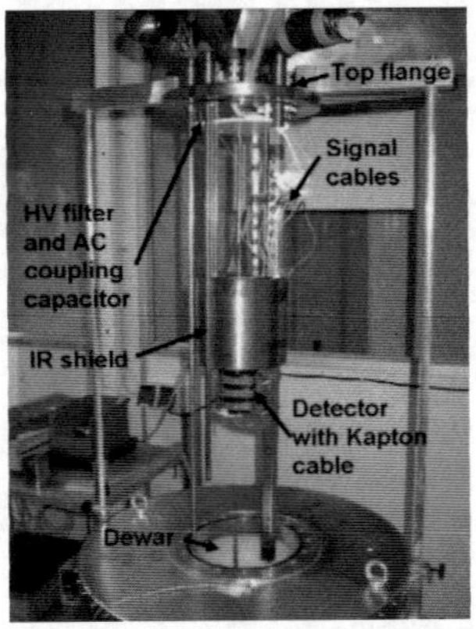

Figure 12. Gerdalinchen II setup.

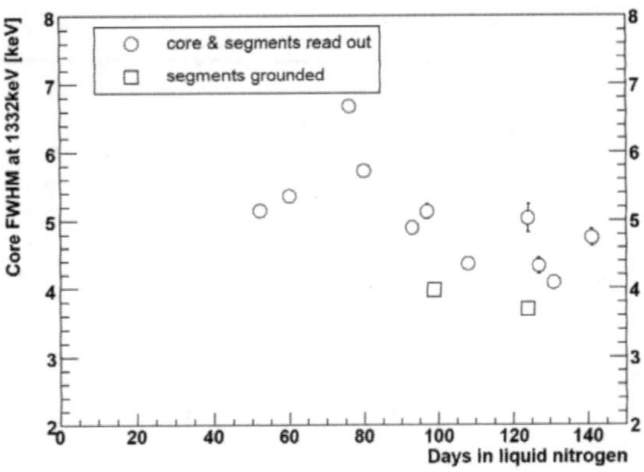

Figure 13. Results of a 140 day test of energy resolution stability of a detector in liquid nitrogen [6].

The results of a test as shown in figure 13 and detailed in [6] have shown that the energy resolution of bare germanium crystals operated over a long period of time is indeed stable. It also verified that the kapton cable snap contacts, as used also for SuSie, are suitable for usage in liquid nitrogen. The facility is available for further tests.

7. Summary and outlook

The GeDet group is concentrating on segmented detectors in order to better understand detector responses, surface effects, charge trapping and charge mobility. These results are used to derive implications for a future ton scale experiment and to establish segmentation as a viable option to reduce background in such an experiment. The group continues to design, build and operate test stands with which to carry out these studies. Alongside this, more general R&D for low background experiments is being carried out in the form of background rejection techniques, and pulse shape simulation and analysis.

References

[1] Agostini M *et al* 2014 *Eur. Phys. J.* C **74** 2764
[2] Abt I, Caldwell A, Doenmez B, Garbini L, Irlbeck S, Majorovits B, Palermo M, Schulz O, Seitz H and Stelzer F 2014 *Preprint* arXiv:1409.0493 [physics.ins-det]
[3] Abt I, Caldwell A, Gutknecht D, Kröninger K, Lampert M, Liu X, Majorovits B, Quirion D, Stelzer F, Wendling P 2007 *Nucl. Instr. Meth. Phys. Res.* A **577** 574-84
[4] Irlbeck S 2014 *The GALATEA test facility and a first study of alpha-induced surface events in a Germanium detector* (München: Ludwig-Maximilians-Universität) p38
[5] Lenz D 2010 *Pulse Shapes and Surface Effects in Segmented Germanium Detectors* (München: Technischen Universität München)
[6] Abt I *et al* 2009 *JINST* **4** p11008

2nd Workshop on Germanium Detectors and Technologies IOP Publishing
Journal of Physics: Conference Series **606** (2015) 012009 doi:10.1088/1742-6596/606/1/012009

Low Background Signal Readout Electronics for the MAJORANA DEMONSTRATOR

I. Guinn[1], N. Abgrall[2], F. T. Avignone III[3,4], A. S. Barabash[5], F. E. Bertrand[4], V. Brudanin[6], M. Busch[7,8], M. Buuck[1], D. Byram[9], A.S. Caldwell[10], Y-D. Chan[2], C. D. Christofferson[10], C. Cuesta[1], J. A. Detwiler[1], Yu. Efremenko[11], H. Ejiri[12], S. R. Elliott[13], A. Galindo-Uribarri[4], G. K. Giovanetti[14,8], J. Goett[13], M. P. Green[4], J. Gruszko[1], V. E. Guiseppe[3], R. Henning[14,8], E. W. Hoppe[15], S. Howard[10], M. A. Howe[14,8], B. R. Jasinski[9], K. J. Keeter[16], M. F. Kidd[17], S. I. Konovalov[5], R. T. Kouzes[15], B. D. LaFerriere[15], J. Leon[1], J. MacMullin[14,8], R. D. Martin[9], S. J. Meijer[14,8], S. Mertens[2], J. L. Orrell[15], C. O'Shaughnessy[14,8], N. R. Overman[15], A. W. P. Poon[2], D. C. Radford[4], J. Rager[14,8], K. Rielage[13], R. G. H. Robertson[1], E. Romero-Romero[11,4], M. C. Ronquest[13], B. Shanks[14,8], M. Shirchenko[6], N. Snyder[9], A. M. Suriano[10], D. Tedeschi[3], J. E. Trimble[14,8], R. L. Varner[4], S. Vasilyev[6], K. Vetter[2] [18], K. Vorren[14,8], B. R. White[4], J. F. Wilkerson[14,8,4], C. Wiseman[3], W. Xu[13], E. Yakushev[6], C-H. Yu[4], and V. Yumatov[5]
The MAJORANA Collaboration

[1]Center for Experimental Nuclear Physics and Astrophysics and Department of Physics, University of Washington, Seattle, WA, USA
[2]Nuclear Science Division, Lawrence Berkeley National Laboratory, Berkeley, CA, USA
[3]Department of Physics and Astronomy, University of South Carolina, Columbia, SC, USA
[4]Oak Ridge National Laboratory, Oak Ridge, TN, USA
[5]Institute for Theoretical and Experimental Physics, Moscow, Russia
[6]Joint Institute for Nuclear Research, Dubna, Russia
[7]Department of Physics, Duke University, Durham, NC, USA
[8]Triangle Universities Nuclear Laboratory, Durham, NC, USA
[9]Department of Physics, University of South Dakota, Vermillion, SD, USA
[10]South Dakota School of Mines and Technology, Rapid City, SD, USA
[11]Department of Physics and Astronomy, University of Tennessee, Knoxville, TN, USA
[12]Research Center for Nuclear Physics and Department of Physics, Osaka University, Ibaraki, Osaka, Japan
[13]Los Alamos National Laboratory, Los Alamos, NM, USA
[14]Department of Physics and Astronomy, University of North Carolina, Chapel Hill, NC, USA
[15]Pacific Northwest National Laboratory, Richland, WA, USA
[16]Department of Physics, Black Hills State University, Spearfish, SD, USA
[17]Tennessee Tech University, Cookeville, TN, USA

E-mail: iguinn@uw.edu

[18] Alternate Address: Department of Nuclear Engineering, University of California, Berkeley, CA, USA

Journal of Physics: Conference Series **606** (2015) 012009 doi:10.1088/1742-6596/606/1/012009

Abstract. The MAJORANA DEMONSTRATOR is a planned 40 kg array of Germanium detectors intended to demonstrate the feasibility of constructing a tonne-scale experiment that will seek neutrinoless double beta decay ($0\nu\beta\beta$) in ^{76}Ge. Such an experiment would require backgrounds of less than 1 count/tonne-year in the 4 keV region of interest around the 2039 keV Q-value of the $\beta\beta$ decay. Designing low-noise electronics, which must be placed in close proximity to the detectors, presents a challenge to reaching this background target. This paper will discuss the MAJORANA collaboration's solutions to some of these challenges.

1. Introduction to the MAJORANA DEMONSTRATOR

The MAJORANA DEMONSTRATOR (MJD) is an array of p-type point contact (PPC) high purity Germanium (HPGe) detectors intended to search for neutrinoless double beta decay ($0\nu\beta\beta$ decay) in ^{76}Ge [1][2][3][4]. MJD will consist of 40 kg of detectors, 30 kg of which will be isotopically enriched to 87% ^{76}Ge. The array will consist of 14 strings of four or five detectors placed in two separate cryostats. One of the main goals of the experiment is to demonstrate the feasibility of building a tonne-scale array of detectors to search for $0\nu\beta\beta$ decay with a much higher sensitivity. This involves acheiving backgrounds in the 4 keV region of interest (ROI) around the 2039 keV Q-value of the $\beta\beta$ decay of less than 1 count/ROI-t-y. Because many backgrounds will not directly scale with detector mass, the specific background goal of MJD is less than 3 counts/ROI-t-y.

MJD uses a wide variety of background reduction techniques. The PPC geometry allows descrimination between multi-site events, consisting mostly of Compton-scatterred gamma backgrounds, and single-site events, which includes $0\nu\beta\beta$ decay. The array is housed in passive shielding of copper, lead and high density polyethylene, along with an active muon veto system. Furthermore, the experiment is located 4850 ft underground at the Sanford Underground Research Facility (SURF), with 4260 mwe overburden to avoid cosmic rays. All materials used inside of the shielding are made of highly radiopure materials; in particular, the copper parts are made out of ultra-pure electroformed copper (EFCu) that is grown underground at SURF. An extensive radio-assay campaign verifies the purity of all materials used in the experiment. This assay data is used in a detailed model of the expected backgrounds of the DEMONSTRATOR[5]. In October 2014, the upper limit on background projected by the model was 3.1 counts in the ROI. A breakdown of the contributions from each background source in the ROI is shown in figure 1, with the largest contributions being the cables, electrical connectors and front end electronics inside of the cryostat. These background predictions are expected to shrink as assay limits improve.

2. Signal Readout Electronics

The signal readout electronics chain is responsible for integrating the current pulses from the Germanium detectors, amplifying that signal and carrying it out of the shielding and to a digitizer. The first stage of integration is done directly above each detector by low-noise, low-mass front-end boards (LMFEs). The signal is carried by bundles of coaxial cables along the strings of detectors and the thermosyphon crossarm to a feedthrough flange, where it is fed into a preamplifier. Because each end of the cable bundles are not easily accessible, signal connectors are placed above the cold plate to allow connection and disconnection of cables. Since all of these components are inside of the cryostat, it is important to design them with radiopure materials and with very low masses to minimize backgrounds, without compromising noise characteristics. The components must further be robust under vacuum and thermal cycling to liquid nitrogen temperature, and must not break when handled inside of a glove box with reduced dexterity.

Journal of Physics: Conference Series **606** (2015) 012009 doi:10.1088/1742-6596/606/1/012009

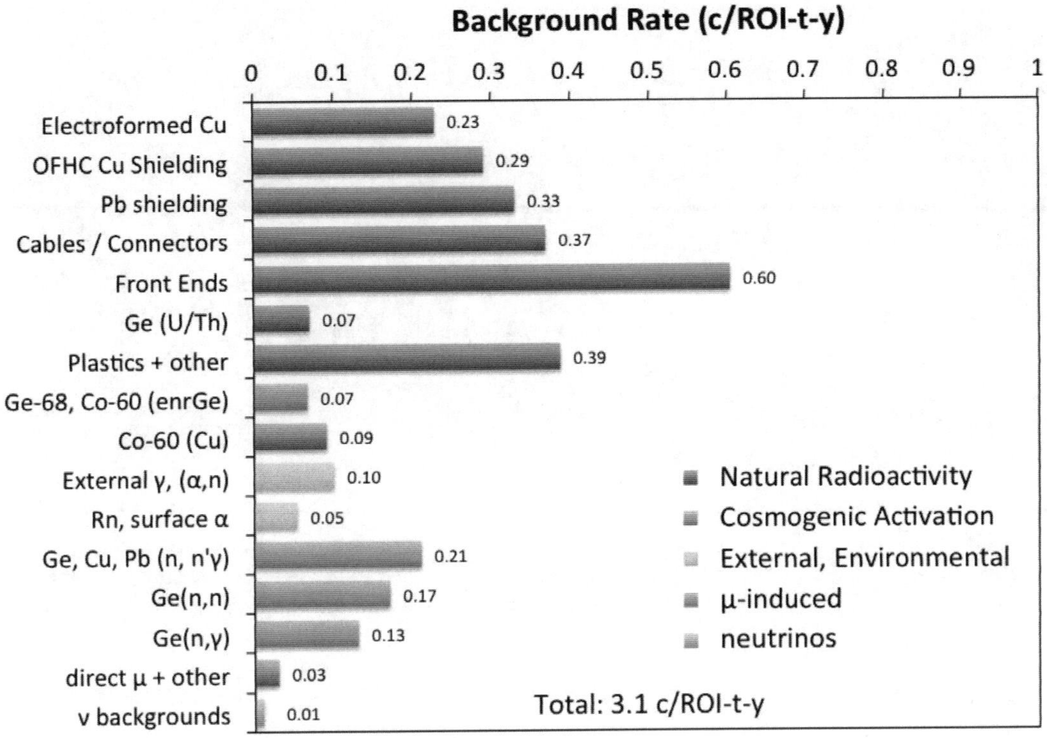

Figure 1. A summary of predicted background for the MAJORANA DEMONSTRATOR, produced in September 2014. The net backgrounds are 3.1 counts/ROI-t-y, with some of the largest contributions coming from signal readout electronics.

2.1. Low-Mass Front-End Boards

The LMFEs, shown in figure 2.1 are resistive feedback charge sensitive amplifiers that act as the first stage of amplification for the detectors [6]. The circuit consists of gold traces on a titanium adhesion layer printed via photolithography on a fused silica substrate. The feedback resistor consists of amorphous Germanium, which has much lower background than more commonly used ceramic resistors. The capacitance is provided by the geometric configuration of the traces. A JFET is attached to the circuit using low background silver epoxy, with the drain and source pads wirebonded to the circuit traces. The circuit also has a pulser line that allows external charge injection that can be used to characterize the circuit. The circuit is housed in an EFCu spring clip, which, when tensioned, presses the circuit against a contact pin at the p^+ point contact of the detector. An ICP-MS assay of an LMFE board and spring clip found backgrounds from Uranium and Thorium at 1.4 μBq/LMFE. The net background from the front-end boards is expected to be 0.60 counts/ROI-t-y.

2.2. Signal Cables

A bundle of four coaxial cables connects each LMFE board to a feedthrough flange. In order to reduce backgrounds, coaxial cables with an outer diameter of 0.4 mm and a mass density of 0.4 g/m are used, custom-made by Axon' Cable S.A.S. in France. A photograph of a bundle of 4 cables is shown, with a BNC coaxial cable for size comparison, in figure 3. Due to this small

Journal of Physics: Conference Series **606** (2015) 012009 doi:10.1088/1742-6596/606/1/012009

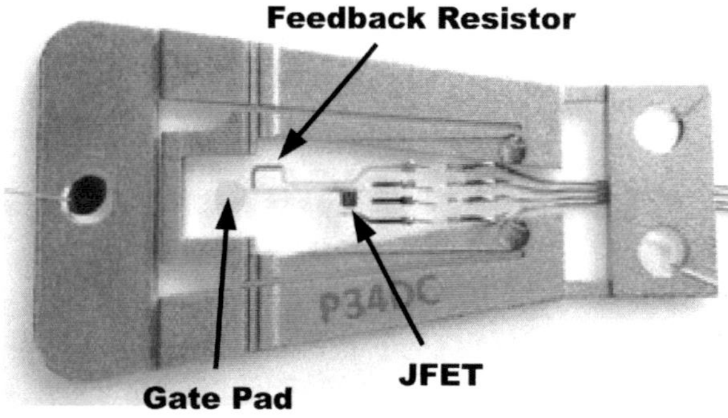

Figure 2. Left: An LMFE board mounted on a spring clip.
Right: an LMFE mounted on a detector unit. The contact pin is pressed between the gate pad and the point contact of the detector.

size, commercial clean copper is used rather than EFCu without heavily impacting backgrounds. The cables have an impedance of 50 Ω over up to 85 in. ICP-MS assay of a short length of cable has found Uranium and Thorium backgrounds of 0.059 µBq/m, providing a total of 0.085 counts/ROI-t-y.

2.3. Signal Connectors
Signal connectors were built to connect two signal bundles together. Because most electrical connectors use electrical contact springs containing Beryllium Copper, which are difficult to make radiopure, no commercially available connectors have low enough backgrounds for MJD. To avoid use of springs, connectors were built that use Mill-Max® gold-plated brass pins with the contact springs removed. The pins and sockets are misaligned slightly, forcing the pin to bend. The restoring force from the bending pin does the job of the spring, maintaining a strong electrical connection. The pins are held in a Vespel® housing, which is machined underground. The signal cables are soldered, using a low background tin-silver eutectic, to the pins and sockets. PTFE Shrink tubing provides strain relief for the cables. Because of the high precision machining needed to properly align the pins and sockets, extensive quality control is necessary to ensure strong connections. A photograph of a signal plug pair is shown in figure 3. ICP-MS assay of the materials in the connectors has found Uranium and Thorium backgrounds at 1.45 µBq/connector, or 0.18 µBq/connection. The total contribution to backgrounds from these connectors is expected to be less than 0.28 counts/ROI-t-y. Using pins with BeCu would increase the background contribution from U and Th to above 10 counts/ROI-t-y.

3. Acknowledgements
This material is based upon work supported by the U.S. Department of Energy, Office of Science, Office of Nuclear Physics. We acknowledge support from the Particle Astrophysics Program of the National Science Foundation. This research uses these US DOE Office of Science User Facilities: the National Energy Research Scientific Computing Center and the Oak Ridge Leadership Computing Facility. We acknowledge support from the Russian Foundation for Basic Research. We thank our hosts and colleagues at the Sanford Underground Research Facility for their support.

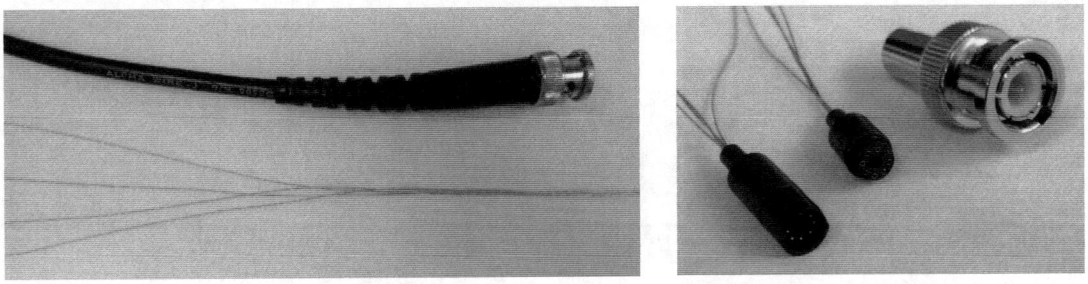

Figure 3. Left: A cable bundle with a BNC cable for comparison.
Right: A male-female signal connector pair, with a BNC connector for comparison.

References

[1] N. Abgrall *et al.*, 2014 The MAJORANA DEMONSTRATOR Neutrinoless Double Beta Decay Experiment *Adv. High Energy Phys.* 2014:365432

[2] F.T. Avignone III, S.R. Elliot, and J. Engel, 2008 Double beta decay, Majorana neutrinos, and neutrino mass *Rev. Mod. Phys.*, 80 481

[3] P.S. Barbeau, J.I. Collar, and O. Tench, 2007 Large-Mass Ultra-Low Noise Germanium Detectors: Performance and Applications in Neutrino and Astroparticle Physics *Journal of Cosmology and Astroparticle Physics*, 09 009

[4] W. Xu *et al.*, 2015 The MAJORANA DEMONSTRATOR: A Search for Neutrinoless Double-beta Decay of ^{76}Ge, Proceedings for Workshop on Germanium-based Detectors and Technologies, 2014 IOP

[5] C. Cuesta *et al.*, 2014 Background Model of the MAJORANA DEMONSTRATOR *Preprint* arXiv:1405.1370

[6] P. Barton *et al.*, 2011 Low-noise low-mass front end electronics for low-background physics experiments using germanium detectors, proceedings for Nuclear Science Symposium and Medical Imaging Conference (NSS/MIC), 2011 IEEE

Journal of Physics: Conference Series **606** (2015) 012010 doi:10.1088/1742-6596/606/1/012010

Coherent elastic neutrino-nucleus scattering

Kate Scholberg

Department of Physics, Duke University, Durham, NC, 27708, USA

E-mail: `schol@phy.duke.edu`

Abstract. I describe physics potential and experimental prospects for coherent elastic neutrino-nucleus scattering (CEvNS), a process which has not yet been observed. Germanium-based detectors represent a promising technology for CEvNS experiments. I focus primarily on stopped-pion neutrino sources.

1. Coherent elastic neutrino-nucleus scattering

Coherent elastic neutrino-nucleus scattering ("CEvNS"[1]) is a process involving the neutral-current scattering of a neutrino with an entire nucleus [1, 2]. It has a well-defined rate in the standard model, given by (for a spin-zero nucleus and neglecting radiative corrections):

$$\frac{d\sigma}{dT}(E,T) = \frac{G_F^2}{2\pi} M \left[2 - \frac{2T}{E} + \left(\frac{T}{E}\right)^2 - \frac{MT}{E^2} \right] \frac{Q_W^2}{4} F^2(Q^2) \,. \tag{1}$$

where E is the incident neutrino energy, T is the nuclear recoil energy, M is the nuclear mass, F is the ground-state elastic form factor, Q_w is the weak nuclear charge, and G_F is the Fermi constant. For nuclei in the medium-A range, the process remains largely coherent (i.e., all nucleon wavefunctions in phase) for neutrino energies up to a few tens of MeV. The cross section is relatively high, boosted by an $\sim N^2$ enhancement, compared to other charged- and neutral-current processes in the same energy range. It potentially even has practical applications [3, 4], thanks to the large cross section.

Although observation of this process has been sought for decades, the experimental signature— very low energy nuclear recoils in the tens-of-keV or less regime— is a challenging one. Only with recent detector developments has good sensitivity to very-low-energy nuclear recoils been feasible. In particular, the experimental search for WIMP dark matter, which has a similar nuclear recoil signature (and in fact for which neutrinos represent a background floor [5, 6]), has driven detector development to enable CEvNS observation.

[1] Note there exist a number of abbreviations for this process in the literature, e.g., CNS, CNNS, CENNS. I favor a version with "E" for "elastic" to distinguish the process from inelastic coherent pion production, which is commonly confused with CEvNS by members of the high energy physics community. I prefer to replace the first "N" with "v", for "neutrino", for two reasons: first, "NN" means "nucleon-nucleon" to many in the nuclear physics community. Second, this disambiguates from CENNS, which is the name of an experimental collaboration. Finally, the Roman letter "v" is less cumbersome than the Greek letter "ν".

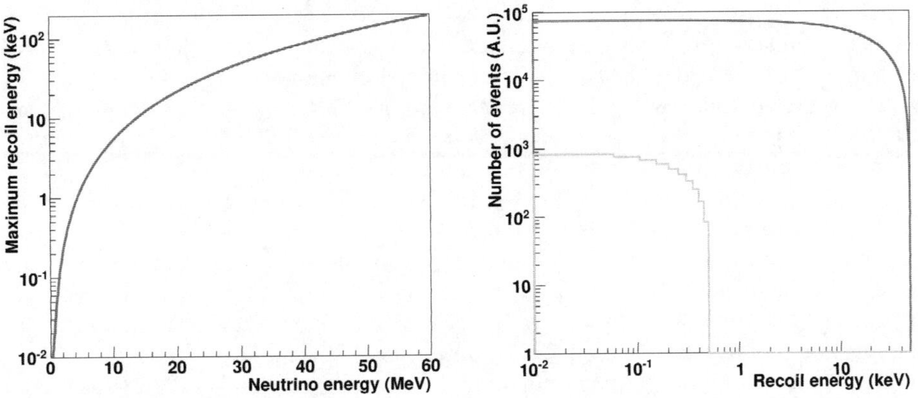

Figure 1. Left: maximum recoil energy for a 30-MeV neutrino coherently scattering from ^{76}Ge. Right: nuclear recoil spectra for 30-MeV (top, red) and 3-MeV (bottom, green) incident neutrinos scattering on ^{76}Ge.

2. Sources and detectors

The CEvNS cross section increases with energy, so in general to detect CEvNS, one wants a relatively high-energy neutrino source. Moreover, the maximum nuclear recoil energy is given by $E_{\rm max} = \frac{2E_\nu^2}{M}$ (see Fig. 1), and the higher the recoil energy of the nucleus, the easier it is in general to detect. Figure 1 shows an example recoil spectrum for 3 and 30 MeV neutrinos, given the same flux, demonstrating both the higher rate and higher recoil energies for higher neutrino energy. However, the neutrino incident energy must not be too high, or else inelastic interactions will begin to dominate (i.e., neutrinos will scatter off individual nucleons inside the nucleus, rather than with the nucleus as a whole— the neutrino wavelength shrinks with increasing energy, and the coherence condition is $Q \sim< 1/R$). For germanium, neutrino energies must be less than about 50 MeV to ensure coherence.

Desirables for a neutrino source for CEvNS detection include:

- High flux
- Well-understood spectrum
- Multiple flavors (helpful for testing new physics)
- A pulsed source if possible, for background rejection
- Ability to get close to the source; flux increases as inverse square of distance
- Practical aspects, such as access and control.

Natural sources of neutrinos suitable for CEvNS detection exist: a nearby core-collapse supernova will create an intense burst, feasibly visible over background, in dark matter detectors (e.g., [7]). Diffuse supernova background neutrinos, solar neutrinos, and the low-energy end of the atmospheric neutrino flux are also in principle observable (geoneutrinos are in principle observable also, but their energies are so low as to produce extremely low-energy recoils). However such steady-state fluxes of natural neutrinos are relatively small and would require enormous detectors, and subject to large backgrounds (perhaps including dark matter!)

Supernova burst neutrinos come in a high-flux pulse, but unfortunately, not on demand. That leaves us with artificial sources for near-term prospects.

There are several candidates for artificial neutrino sources for CEvNS detection. Reactors are among the most promising, producing enormous fluxes of $\bar{\nu}_e$ (and CEvNS has possible practical use for reactor monitoring [3, 4]). However, reactor neutrino energies range only up to $\sim$7 MeV, presenting the experimentalist with the significant challenge of detecting of mostly sub-keV recoils (see Fig. 1). Furthermore, these are steady-state sources and backgrounds must therefore be kept well under control. Radioactive sources can also be made quite intense; however these have typically even lower energies than reactor neutrinos. Beam-induced radioactive sources, such as IsoDAR [8], are under development; neutrino energies are still lower than 15 MeV and neutrino production steady-state; nevertheless they are interesting. Storage rings of boosted decaying nuclei, the so-called "beta beams", which have tunable energy, could also be very interesting [9]. These, too, are steady-state sources and so require excellent background rejection. These latter two kinds of sources, while possibly interesting for the future, do not yet exist.

The remaining near-term possibility for an artificial neutrino source for CEvNS detection is a pion decay at rest (stopped-pion) source. Such sources have previously been used successfully for neutrino physics [10, 11]. The idea is to irradiate a dense target with a proton beam (preferably with proton energies below kaon production threshold). The resulting pions decay at rest, producing neutrinos of muon, electron, and muon antineutrino flavor with well understood spectra, as shown in Fig. 2. These neutrinos have high enough energy to produce promising-to-detect recoils, yet have energies low enough to result in a high fraction of coherent interactions in most detector targets.

Table 1 contrasts the pros and cons of reactor and stopped-pion neutrinos for CEvNS detection.

Table 1. Properties and pros and cons for CEvNS measurements of reactor and pion decay-at-rest neutrinos.

Source	Flux ν's per second	Flavor	Energy	Pros	Cons
Reactor	2×10^{20} per GW	$\bar{\nu}_e$	few MeV	· huge flux	· lower cross section · low energy recoils · steady-state
Stopped-pion	1×10^{15}	ν_μ ν_e $\bar{\nu}_\mu$	0-50 MeV	· higher cross section · higher energy recoils · pulsed beam · multiple flavors	· lower flux · in-time neutrons

Table 2 summarizes properties of past, existing and future stopped-pion sources worldwide. Figure 4 plots these in a figure-of-merit space. Neutrino flux is roughly proportional to power, so higher power is better. The larger the reciprocal duty factor, the better for background rejection (although beam time windows smaller than the muon decay timescale provide only marginal improvement for the delayed flux).

Of these, the Spallation Neutron Source (SNS) in Oak Ridge, TN, provides an excellent prospect [15]. The SNS uses a proton beam energy of about 1 GeV incident on a liquid mercury target, for total power up to 1.4 MW, and resulting in about 8×10^{14} neutrinos per flavor per second produced at the source. The repetition rate is 60 Hz, and pulse width is about 700 ns, resulting in a background rejection factor of 10^{-3} to 10^{-4}. Because the proton energy is not

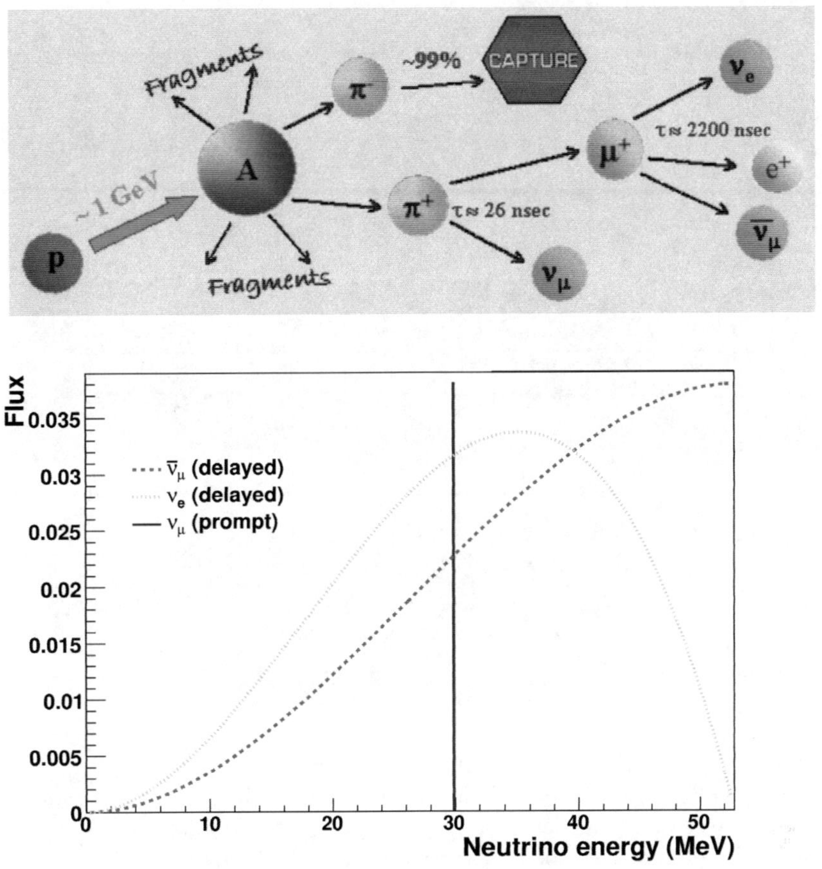

Figure 2. Top: SNS neutrino production mechanism [12]. Bottom: resulting neutrino spectrum (flux in arbitrary units) from pion decay at rest.

Table 2. Characteristics of past, current and planned stopped-pion neutrino sources worldwide.

Facility	Location	Proton Energy (GeV)	Power (MW)	Bunch Structure	Rate (Hz)
LAMPF	USA (LANL)	0.8	0.8	600 μs	120
ISIS	UK (RAL)	0.8	0.16	2 $\times$ 200 ns	50
BNB	USA (FNAL)	8	0.032	1.6 μs	5-11
SNS	USA (ORNL)	1.3	1.4	700 ns	60
MLF	Japan (J-PARC)	3	1	2 $\times$ 60-100 ns	25
CSNS	China (planned)	1.6	0.1	<500 ns	25
ESS	Sweden (planned)	1.3	5	2 ms	17
DAEδALUS	TBD (planned)	0.7	$\sim 7 \times 1$	100 ms	2

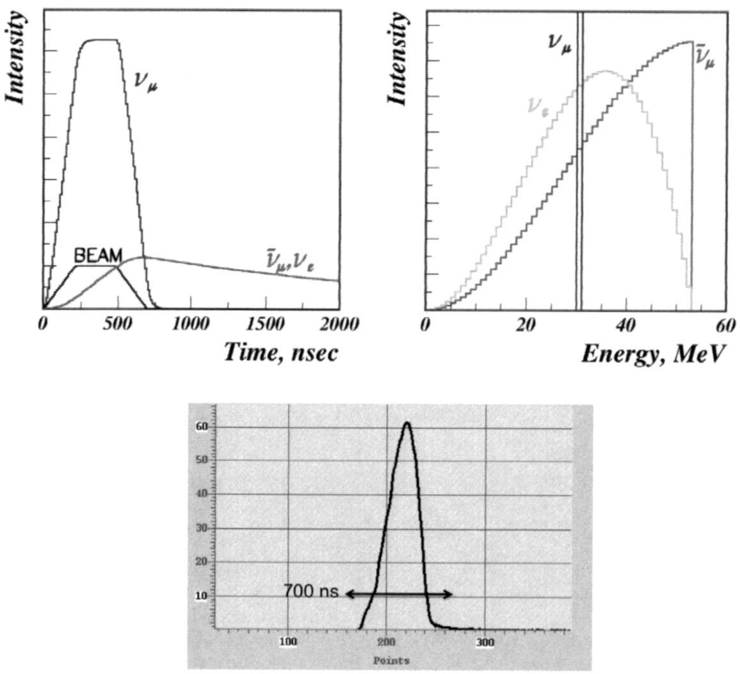

Figure 3. Time and energy distributions (arbitrary units) for the different neutrino flavors produced at the SNS. The top plots are from a 2003 study [13]; the bottom plot shows proton beam structure on target from recent SNS running [14]. The prompt (pion-decay) component of the neutrino flux should closely follow the proton time structure.

too high (below kaon production threshold), the contamination from non-DAR-pion neutrinos should be small. This neutrino source is of extremely high quality.

Expected event rates for CEvNS in various targets are shown in Fig. 5. The effect of kinematics is seen here: the heavier the target, the more events are expected, but fewer are kicked to high recoil energy. Germanium rates are promising for detection thresholds less than a few tens of keV.

3. Physics sensitivity

The cross section for CEvNS is very cleanly predicted in the standard model; hence, a deviation could be the signature of beyond-the-SM physics. Because of the dependence of the cross section on the weak mixing angle, in the context of the standard model, it represents a new channel for measuring $\sin^2\theta_W$— so a deviation from prediction would mean potentially new physics. The weak mixing angle measurement in a first-generation experiment would not likely have better precision than other current experiments [16]. However, certain new physics that is not probed by the existing weak mixing angle measurements — in particular, non-standard interactions (NSI) of neutrinos with quarks — would result in an anomalous interaction rates. Even a first-generation measurement would provide constraints an order of magnitude better than existing ones [17, 18, 19]. Multiple types of targets, with different N and Z, will be advantageous for cancelling systematics. Other possibilities are neutrino magnetic moment searches [17]. In

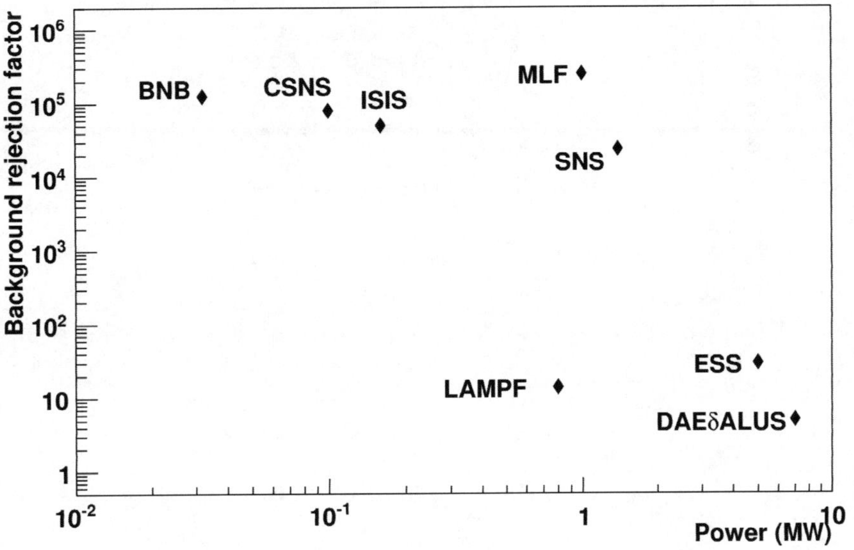

Figure 4. Figure-of-merit space for pion decay-at-rest neutrino sources. The x-axis shows power; neutrino flux is approximately proportional to power, so one wants to be further to the right. The y-axis is the reciprocal duty factor, $1/(TR)$, where T is the beam pulse window length and R is the pulse rate. This is equivalent to the background rejection factor for steady-state backgrounds; one wants this factor to be as large as possible.

the longer term, after experimental precision exceeds that of the current knowledge of nuclear form factors [20, 21, 22], a CEvNS measurement can in principle probe these nuclear form factors via precision spectral shape measurements. Another potentially interesting far-future prospect is the sensitivity to sterile neutrino oscillations; because the channel is neutral current, and hence flavor-blind, a disappearance experiment could probe parameter-space regions of current interest [23]. Note that because the CEvNS process represents a background "floor" for dark matter detection experiments [5, 6], it is essential to measure it to confirm predictions. The CEvNS process is astrophysically important, playing a role in supernova processes and detection [7], and may even potentially be used for solar neutrino physics [24].

A first measurement could in principle be made at a pion decay-at-rest neutrino source with a few to a few tens of kilograms of detector material. For such a "first light" experiment, precision understanding of flux and stringent control of systematic uncertainties is not needed. A next-phase experiment, up to a few hundred kilograms, could provide new constraints on NSI of neutrinos with quarks, and will start to become limited by systematic uncertainties. An eventual tonne- or multi-tonne-scale experiment could probe nuclear physics, and with low enough energy threshold, also potentially search for large neutrino magnetic moment. Exacting control of systematics will be required at this stage.

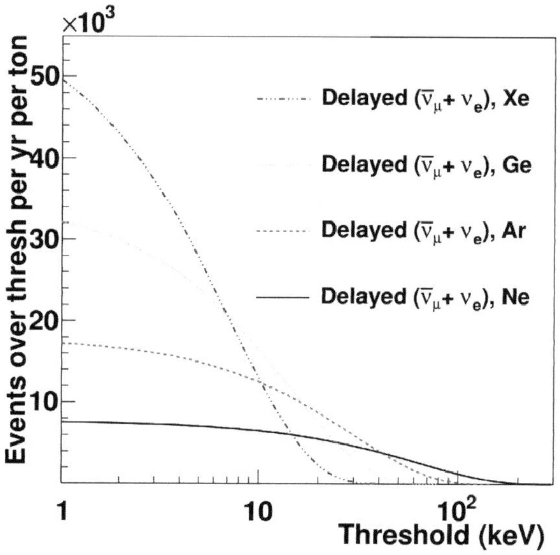

Figure 5. CEvNS events from the muon-decay component of the pion decay-at-rest flux over recoil energy threshold for different materials for a location 20 m from the SNS target. The rates are for spin-zero isotopes ^{20}Ne, ^{40}Ar, ^{76}Ge, ^{132}Xe, but should be similar for other isotopes.

4. Prospects for near-future measurements

While there are several very interesting proposals for measuring CEvNS at reactors with various technologies, including germanium-based (e.g.,[25, 26, 27]), I will focus here on proposals to use pion decay-at-rest neutrinos.

4.1. COHERENT

The COHERENT collaboration is newly formed and aims to use the SNS at Oak Ridge National Laboratory [28]. The aim is to make a CEvNS measurement with as many different detector targets as resources permit. Three technologies are currently under consideration: two-phase liquid xenon, CsI[Na], and germanium. The germanium-based detectors under consideration are high-purity PPC technology, deployed as MAJORANA Demonstrator-like strings [29]: see Figs. 6 and 7. An initial deployment of a few kg could be increased to a few tens of kg. Fig. 8 shows the expected differential recoil spectrum.

Neutron-induced nuclear recoils are the most significant background. While cosmic neutrons and other cosmogenic and ambient backgrounds can be well understood from data outside of the beam pulse window and subtracted, beam-related neutrons are more worrisome. A background campaign over 2013-2014 using scintillator neutron detectors has found some promising sites at the SNS, including one with some overburden in the basement. Neutrino-induced neutrons from charged- and neutral-current interactions in a lead shield surrounding the detectors may cause some non-negligible background [30].[2]

[2] These neutrino-induced neutrons are interesting in themselves, for e.g., supernova detection [31], and measurements are underway.

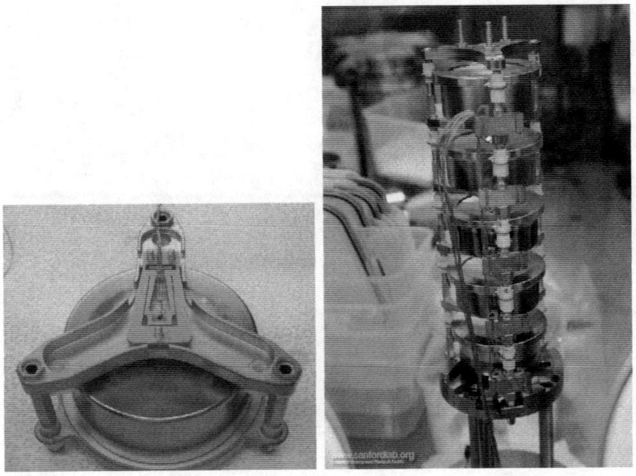

Figure 6. Canberra Broad-Energy Germanium (BEGe) PPC-type detectors in the Majorana Demonstrator detector mounting hardware. Left: a single mounted detector with low-mass front end electronics. Right: A string of 5 detectors with total mass of ~ 3 kg (Photo credit: Matthew Kapust, Sanford Underground Research Facility). From reference [28].

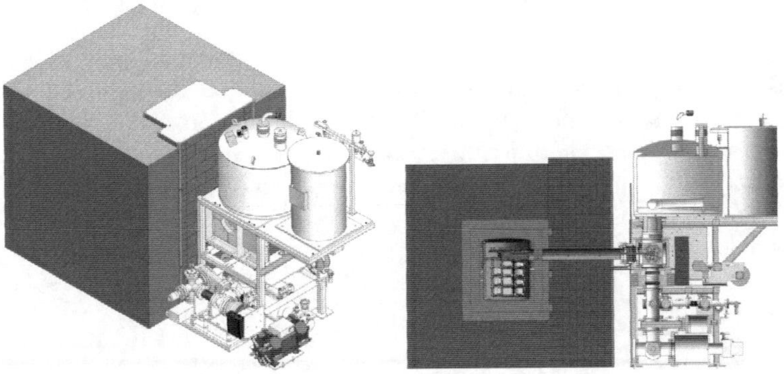

Figure 7. A Majorana Demonstrator modular cryostat with capacity of ~20 kg of germanium detectors, deployed in a potential passive shield of copper and lead. From reference [28].

4.2. CENNS
Another collaboration aiming to exploit stopped-pion neutrinos for CEvNS measurement is the CENNS collaboration [32]. Here the idea is to exploit the Fermilab Booster Neutrino Beam flux at a far off-axis location, where the flux is dominated by pion decay-at-rest neutrinos. The detector concept is a single-phase argon detector in a water shield, similar to the proposed CLEAR detector [33]. Background measurements using the SciBath detector [34] have characterized the beam-related neutron background, to determine required levels of shielding for a measurement. The background rejection factor is quite good, although flux is relatively low (see Fig. 4).

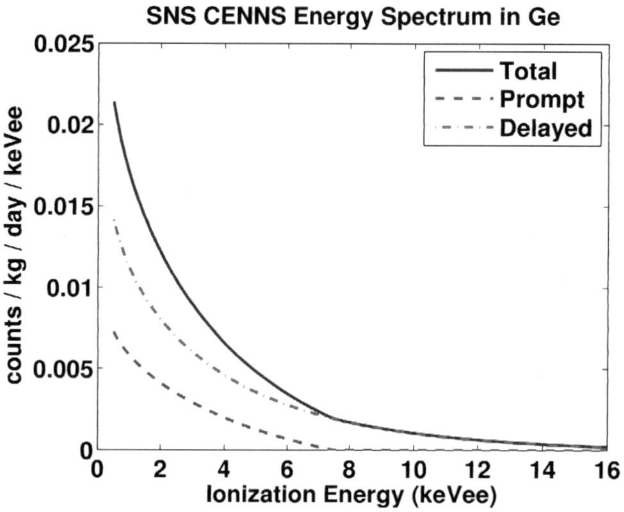

Figure 8. CENNS ionization-energy (keVee) spectrum (events per keVee per kg active mass of Ge per day of observation time) in a Ge detector located 20 m from the SNS target (from reference [28]). The prompt and delayed components of the spectrum are shown separately.

5. Summary

In summary, CEvNS measurements offer a number of interesting physics prospects. The low-hanging fruit is NSI, for which significant constraints can be set even in a first-generation program. Future-generation experiments will have a broad physics program. For the first measurements, requirements are not terribly stringent; however in the long term, multiple targets and well understood systematics will be necessary. Pion decay at rest sources are particularly attractive, thanks to high cross sections and recoil energies, as well as good background rejection. Reactor sources are attractive for high fluxes. Many interesting detector strategies have been proposed, including germanium, such as for the COHERENT proposal. Prospects are good for near-term measurements.

References

[1] Freedman D Z, Schramm D N and Tubbs D L 1977 *Ann.Rev.Nucl.Part.Sci.* **27** 167–207
[2] Drukier A and Stodolsky L 1984 *Phys.Rev.* **D30** 2295
[3] Barbeau P, Collar J, Miyamoto J and Shipsey I 2003 *IEEE Trans. Nucl. Sci.* **50** 1285–1289 (*Preprint* `hep-ex/0212034`)
[4] Hagmann C and Bernstein A 2004 *IEEE Trans.Nucl.Sci.* **51** 2151–2155 (*Preprint* `nucl-ex/0411004`)
[5] Monroe J and Fisher P 2007 *Phys.Rev.* **D76** 033007 (*Preprint* `0706.3019`)
[6] Billard J, Strigari L and Figueroa-Feliciano E 2013 (*Preprint* `1307.5458`)
[7] Horowitz C, Coakley K and McKinsey D 2003 *Physical Review D* **68** 23005
[8] Bungau A, Adelmann A, Alonso J, Barletta W, Barlow R *et al.* 2012 *Phys.Rev.Lett.* **109** 141802 (*Preprint* `1205.4419`)
[9] Volpe C 2004 *J.Phys.* **G30** L1–L6 (*Preprint* `hep-ph/0303222`)
[10] Zeitnitz B (KARMEN) 1994 *Prog.Part.Nucl.Phys.* **32** 351–373
[11] Athanassopoulos C *et al.* (LSND) 1997 *Nucl.Instrum.Meth.* **A388** 149–172 (*Preprint* `nucl-ex/9605002`)
[12] Nu-SNS Collaboration 2005 URL `http://www.phy.ornl.gov/nusns/proposal.pdf`
[13] Avignone F and Efremenko Y 2003 *J.Phys.* **G29** 2615–2628
[14] Galambos J SNS High Power Proton Beam Experience and Plans Talk at COHERENT meeting
[15] Bolozdynya A, Cavanna F, Efremenko Y, Garvey G, Gudkov V *et al.* 2012 (*Preprint* `1211.5199`)

[16] Olive K *et al.* (Particle Data Group) 2014 *Chin.Phys.* **C38** 090001
[17] Scholberg K 2006 *Phys.Rev.* **D73** 033005 (*Preprint* hep-ex/0511042)
[18] Barranco J, Miranda O and Rashba T 2005 *JHEP* **0512** 021 (*Preprint* hep-ph/0508299)
[19] Barranco J, Miranda O and Rashba T 2007 *Phys.Rev.* **D76** 073008 (*Preprint* hep-ph/0702175)
[20] Amanik P and McLaughlin G 2009 *J.Phys.G* **G36** 015105
[21] Patton K, Engel J, McLaughlin G and Schunck N 2012 (*Preprint* 1207.0693)
[22] Patton K, McLaughlin G and Scholberg K 2013 *Int.J.Mod.Phys.* **E22** 1330013
[23] Anderson A, Conrad J, Figueroa-Feliciano E, Ignarra C, Karagiorgi G *et al.* 2012 *Phys.Rev.* **D86** 013004 (*Preprint* 1201.3805)
[24] Billard J, Figueroa-Feliciano E and Strigari L 2014 (*Preprint* 1409.0050)
[25] Wong H T 2008 *Mod.Phys.Lett.* **A23** 1431–1442 (*Preprint* 0803.0033)
[26] Collar J (CoGeNT Collaboration) 2008 *J.Phys.Conf.Ser.* **136** 022009
[27] Sangiorgio S, Bernstein A, Coleman J, Foxe M, Hagmann C *et al.* 2012 *Phys.Procedia* **37** 1266–1272
[28] J A *et al.* 2014 https://fsnutown.phy.ornl.gov/fsnufiles/positionpapers/Coherent_PositionPaper.pdf
[29] Abgrall N *et al.* (Majorana Collaboration) 2014 *Adv.High Energy Phys.* **2014** 365432 (*Preprint* 1308.1633)
[30] Collar J, Fields N, Fuller E, Hai M, Hossbach T *et al.* 2014 (*Preprint* 1407.7524)
[31] Vaananen D and Volpe C 2011 *JCAP* **1110** 019 (*Preprint* 1105.6225)
[32] Brice S, Cooper R, DeJongh F, Empl A, Garrison L *et al.* 2014 *Phys.Rev.* **D89** 072004 (*Preprint* 1311.5958)
[33] Scholberg K, Wongjirad T, Hungerford E, Empl A, Markoff D *et al.* 2009 (*Preprint* 0910.1989)
[34] Cooper R, Garrison L, Meyer H O, Mikev T, Rebenitsch L *et al.* 2011 (*Preprint* 1110.4432)

Journal of Physics: Conference Series **606** (2015) 012011 doi:10.1088/1742-6596/606/1/012011

Germanium detectors with sub-keV sensitivities for neutrino and dark matter physics

Arun Kumar Soma and Henry Tsz-King Wong
(On behalf of the TEXONO Collaboration)
Institute of Physics, Academia Sinica, Taipei 11529, Taiwan
E-mail: arunsoma@phys.sinica.edu.tw, htwong@phys.sinica.edu.tw

Abstract. A detector of $\mathcal{O}(1$ kg) modular mass with $\mathcal{O}(100$ eV) threshold at $\mathcal{O}(1$ kg^{-1}keV^{-1}day$^{-1})$ background level finds tremendous application in the field of neutrino and dark matter physics. This novel detector demands overcoming several challenges at both hardware and software levels. The collaboration is exploring Germanium detection technology and highlights of the R & D program are presented. The salient features of various detector configuration and the applied analysis methodologies are discussed. In particular the differentiation of surface and bulk events by pulse shape analysis in point contact Germanium detector is described. These advances pave the way for new detector technique to be fully exploited.

1. Physics motivations

Germanium (Ge) detectors offer potential for realizing $\mathcal{O}(1$ kg) modular mass, $\mathcal{O}(100$ eV) threshold and $\mathcal{O}(1$ kg^{-1}keV^{-1}day$^{-1})$ background level experiments. When these benchmark specifications are fulfilled, several topics discussed in subsequent sub-sections can be experimentally pursued.

1.1. Neutrino electromagnetic properties

The studies on possible neutrino electromagnetic interactions [1] can probe physics beyond the Standard Model. Neutrino magnetic moments (μ_ν) is an intrinsic neutrino property that describes possible neutrino-photon couplings via its spin [2, 3]. The helicity is flipped in μ_ν induced interactions. Observations of μ_ν at levels relevant to present or future generations of experiments will strongly favor the neutrinos are Majorana particles [4]. As shown in Figure 1a, the μ_ν contributions are enhanced as T decreases, with modifications of the atomic binding energy effects [5]. Where, T denotes the measurable energy in the detector (in this case it originates from the recoil energy of electron).

In a similar spirit, studies of neutrino "milli-charge" (Q_ν) probe possible helicity-conserving QED-like interactions. Finiteness of Q_ν would imply neutrinos are Dirac particles. It was recently recognized that there is enhancement in cross-sections due to atomic effects [6], as depicted in Figure 1a. The known ratios of the peaks and discrete binding energies provide smoking gun signatures for positive observations.

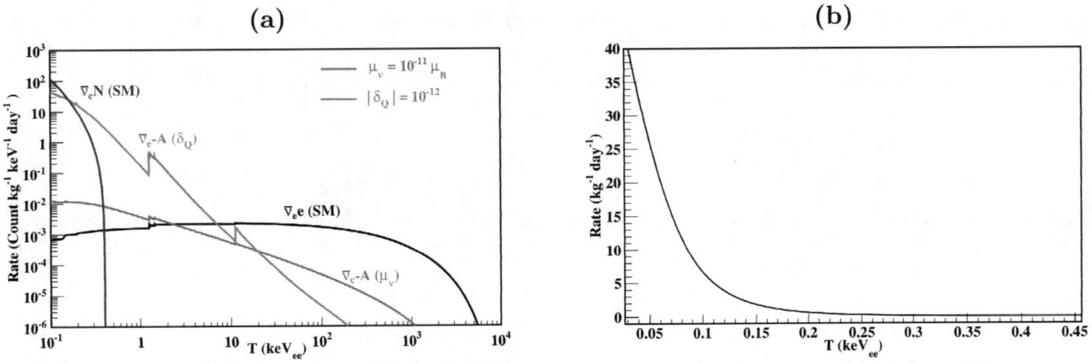

Figure 1. (a) The observable spectra due to neutrino interactions on Ge target with reactor $\bar{\nu}_e$ at $\phi(\bar{\nu}_e)$=10^{13} cm^{-2}s^{-1}. The neutrino magnetic moment (MM) contributions to $\bar{\nu}_e$-e at μ_ν = 10^{-10} μ_B, as well as the Standard Model (SM) $\bar{\nu}_e$-e and coherent scattering $\bar{\nu}_e$-N are shown. (b) Expected integral $\bar{\nu}_e$-N coherent scattering rates due to SM contributions as a function of detector threshold, assuming realistic detector resolution.

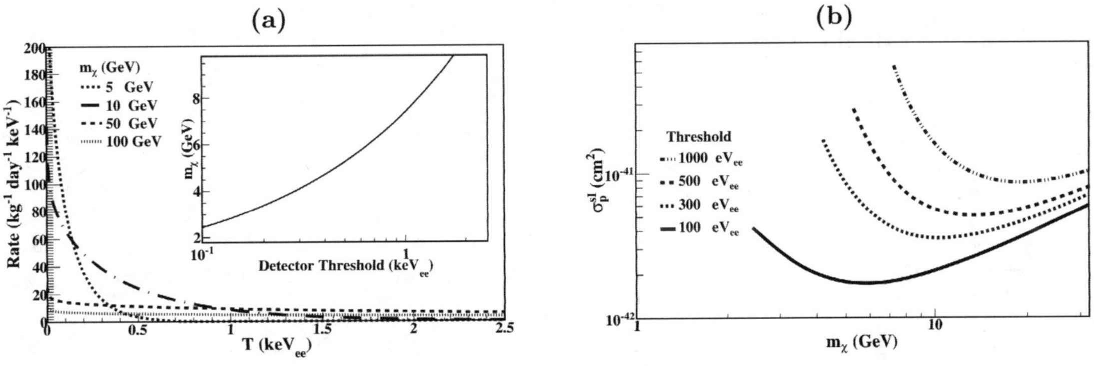

Figure 2. (a) Measurable recoil spectra for WIMP-Ge interactions at a cross-section of 10^{-40} cm^2, at several values of m$_\chi$. The lower bounds of the m$_\chi$ which can be probed with 1 kg-yr of data at background level of 1 kg^{-1}keV^{-1}day^{-1} as a function of physics threshold is shown in the inset. Quenching effects of nuclear recoils are taken into account. (b) Sensitivity reach of the same configuration at different detector threshold, showing the relative improvement in cross-section as a function of m$_\chi$.

1.2. Neutrino nucleus coherent scattering

Neutrino coherent scattering with the nucleus (ν + $N \rightarrow \nu$ + N, νN) [7] is a fundamental neutrino interaction which has never been observed. Measurement of νN coherent scattering would provide a sensitive test to the Standard Model. The coherent interaction plays an important role in astrophysical processes. They may provide new approaches towards the detection of supernova neutrinos and offer a promising avenue towards a compact and relatively transportable neutrino detector for real-time monitoring of nuclear reactors.

The maximum nuclear recoil energy for Ge target (A=72.6) due to reactor $\bar{\nu}_e$ is about 2 keV. The quenching factor is about 20% for Ge in the <10 keV region. Accordingly, the maximum measurable energy is about 400-500 eV [12]. The typical differential spectrum and the integral event rate as a function of measurement threshold are given in Figures 1a&b, respectively. At the benchmark sensitivities, the expected rate is $\mathcal{O}(10\ \ \mathrm{kg^{-1}day^{-1}})$ with signal to background ratio >50 at threshold of 100 $\mathrm{eV_{ee}}$. Low detector threshold is therefore crucial in such experiments.

1.3. Dark matter searches

The WIMP interaction with matter [8] ($\chi\ +\ N\ \rightarrow\ \chi\ +\ N$) is predominantly via the same coherent scattering mechanism like that of neutrinos. The major difference is that cosmology requires the WIMPs to be massive and its motion is non-relativistic to be consistent with observational data on structure formation [8]. In addition, there may be both spin-independent and spin-dependent interactions between WIMP and matter. The nuclear recoil spectra at 10^{-40} cm^2 for several WIMP mass (m_χ) for Ge are displayed in Figure 2a.

Therefore, the advantages of having a low threshold detector are as follows:

(i) WIMPs with lower masses would become detectable when nuclear recoils at lower energy can be measured, thereby opening a new window of observation. The sensitivity reach of m_χ with 1 kg-yr of data strength as a function of threshold at a background of 1 $\mathrm{kg^{-1}keV^{-1}day^{-1}}$ is illustrated in the inset of Figure 2a.

(ii) The low threshold allows a larger range of WIMPs to contribute in an observable interaction and hence results in better sensitivities for all values of m_χ, as illustrated Figure 2b.

2. Experimental configuration

The Ge detection technology is being explored by TEXONO [11] and CDEX [14] experiments at Kuo-Sheng Reactor Neutrino Laboratory (KSNL, 28 meters from a 3 GW thermal power reactor core, 30 meter-water-equivalent overburden) and China Jinping Underground Laboratory (CJPL, >2400 m of rock overburden with drive-in access), respectively. The reasons for exploring the Ge detection technology are as follows:

(i) Ge is a semiconducting material with an band-gap of ~0.66 eV at 300 K and requires ~2.96 eV for an electron-hole pair at 77 K [9]. Therefore, sub-keV threshold detectors can be fabricated.

(ii) Ge detection technology is well matured. Therefore, desired radio-purity and detectors with larger modular mass can be constructed.

2.1. Active and passive shielding

The "baseline design" is shown in the Figure 3. Ge detectors are enclosed by an NaI(Tl) anti-Compton (AC) detector and copper passive shielding inside a plastic bag purged by nitrogen gas evaporated from liquid nitrogen dewar. The set-up is further shielded by copper, boron-loaded polyethylene, steel and lead. At the surface laboratory (KSNL), an additional shielding in the form of cosmic-ray (CR) veto panels comprising of plastic scintillators that are read out by photomultipliers is deployed. The AC and CR detectors are crucial and serve as vetos to reject background and as tags to identify samples for efficiency measurements. The details of signal processing and data analysis can be referred to [11, 14, 15].

2.2. Sub-keV threshold Ge detectors

TEXONO collaboration has studied operational characteristics and performance of various commercially constructed Ge detectors. The four kinds of detector configurations used[12] are listed below with their corresponding sensor schematics depicted in Figure 4.

(i) 1 kg coaxial p-type high purity Ge detector used as target detector in Ref. [3],

(ii) 4×5 g n-type Ge detector array (ULEGe) used as target detector in Refs. [10, 13],

Journal of Physics: Conference Series **606** (2015) 012011 doi:10.1088/1742-6596/606/1/012011

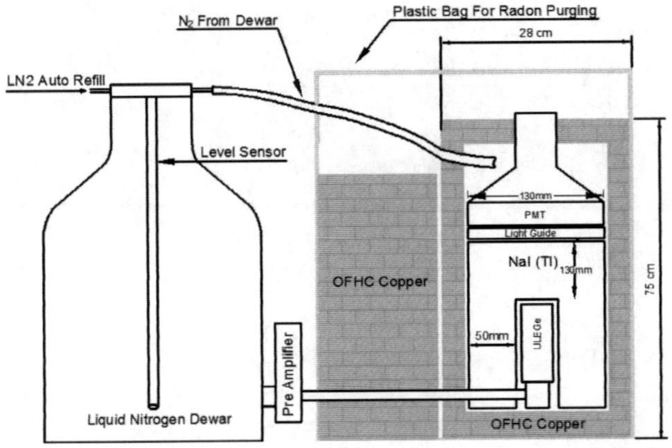

Figure 3. Schematic diagram of experimental set-up that includes Ge and NaI(Tl) scintillator along with 50-tons of passive shielding structure.

(iii) 900 g p-type point-contact Ge detector (pPCGe) used as target detector in Refs. [11, 14] and

(iv) 500 g n-type point-contact Ge detector (nPCGe) used for calibration in Refs. [11, 15].

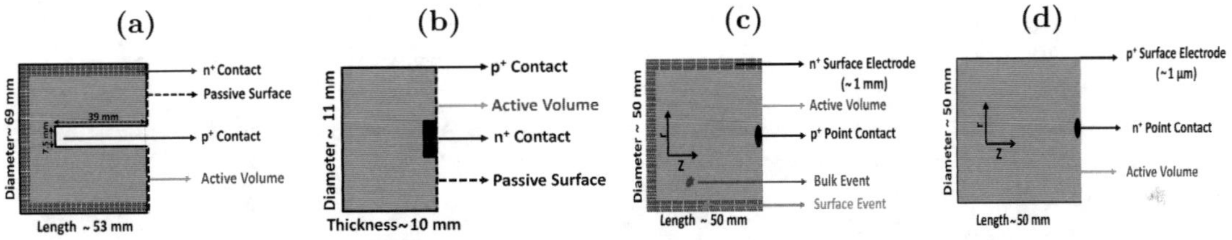

Figure 4. Schematic crystal configurations of the Ge detectors: (a) coaxial HPGe at 1 kg mass (b) ULEGe at 5 g modular mass (c) pPCGe at 900 g mass and (d) nPCGe at 500 g mass.

The HPGe and ULEGe are conventional technologies. The concept of point-contact Ge detectors was proposed and the first nPCGe at 800 g mass was constructed in the 1980's [16], followed by recent realization of pPCGe [17]. The ULEGe, pPCGe and nPCGe are of sub-keV sensitivities and with typically ~pF capacitance.

A summary of detectors performance is given in Table 1. PCGe detectors are the focuses of subsequent discussions and are adopted as "Reference" detector to illustrate characterization processes and evolution of background spectra with software selection procedures.

The crystal of pPCGE is made of p-type Ge. The outer surface electrode is at positive high voltage towards which the electrons are drifted. The central point contact electrode is at zero-potential. The outer surface electrode is fabricated by lithium diffusion and its thickness is ~1 mm. The electron-hole pairs produced by radiations at surface (S) layer are subjected to a weaker drift field than those at the bulk (B) region of crystal. A portion of the pairs will recombine while the residual will induce signals which are weaker and slower than those originated in B. Therefore, S-events have partial charge collection and slower rise-time [15].

The pre-amplifiers are coupled to point and surface contact from which electrical signals are extracted. These signals are then processed by shaping and timing amplifiers before feeding to

Journal of Physics: Conference Series **606** (2015) 012011 doi:10.1088/1742-6596/606/1/012011

Table 1. Summary table on the performance of various Ge detectors.

Items	HPGe	ULEGe	pPCGe	nPCGe
Modular Mass (g)	1000	5	900	500
RESET Amplitude (V)	N/A †	8.0	9.8	6.8
RESET Time Interval (ms)	N/A †	700	800	170
Pedestal Noise				
Amplitude RMS σ_A (eV)	812	15	56	49
Charge(Energy) RMS σ_Q (eV)	840	38	79	52
Pulser Width				
FWHM (eV)	1566	80	141	133
RMS (eV)	665	34	60	57
Gamma Line Width	Ga-K X-Ray	^{55}Fe	Ga-K X-Ray	Ga-K X-Ray
RMS (eV)	910	78	157	134
Selected Trigger Level Δ (σ_A)	3.58	4.3	3.64	3.73
50% Trigger Efficiency (eV)	3500	80	230	200
Trigger Rate (Hz)	60	5	42	36
50% Selection Efficiency (eV)	4200	220	400	300
Noise Edge "NoEd" (eV)	5000	320	550	350

† HPGE uses Resistance feedback preamplifier and so not applicable (N/A).

60 MHz and 200 MHz digitizes, respectively. The process of noise and background suppression are presented in the next section.

The thickness of S layer was measured to be (1.16±0.09)mm by comparing intensity ratios of simulated and observed γ-peaks from ^{133}Ba source [18]. The corresponding fiducial mass for B-region is 840 g.

2.3. Physics event selection

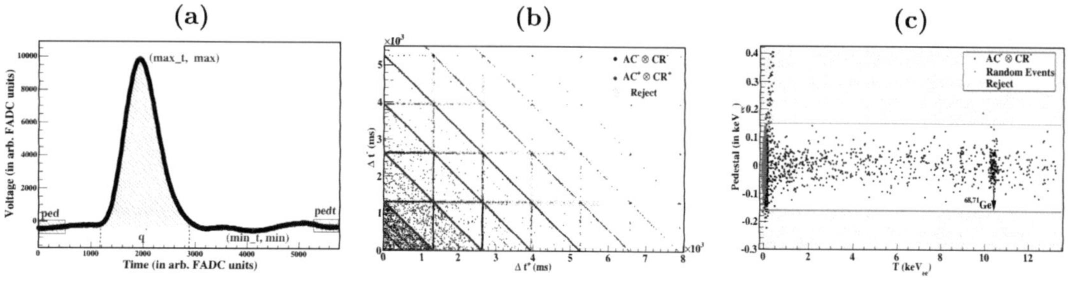

Figure 5. (a) Shaping amplifier (Canberra 2026) output with definitions of some of the analysis parameters; (b) Events timing with respect to reset signal, where Δt^+ and Δt^- are correspondingly the next and previous reset signal; (c) events selection based on fluctuation of pedestal.

A typical shaping amplifier (SA) output pulse is shown in Figure 5a. The waveform is characterized by several parameters and some of them are shown in Figure 5a. The noise events are then eliminated by exploring correlations between the stored waveform parameters. For example, events timing correlated with pre-amplifier reset signal and distribution of pedestal are shown in Figure 5b and 5c, respectively. The cumulative efficiency of all such cuts (PN: Physics and Noise) is shown in Figure 8c.

The energy calibration was performed by using $X - rays$ from ^{68}Ge (10.37 keV), ^{68}Zn (8.98 keV), and ^{68}Ge + ^{65}Zn (1.29 keV) with random trigger events providing zero-energy definition [12].

The nuclear recoil χN and νN events are uncorrelated with other detector components and are uniformly distributed in pPCGe volume. If superscript $-/(+)$ denotes anti-coincidence (coincidence) of cosmic ray and anti-Compton veto systems with pPCGe signals, then desired signal events are designated by $AC^{-} \otimes CR^{-}$. The $AC^{+} \otimes CR^{-}$ and $AC^{-} \otimes CR^{+}$ select ambient gamma and cosmic ray induced high energy neutron events, respectively.

3. Differentiation of bulk and surface events

The typical pulse shape of fast-timing amplifier (TA) are shown in Figure 6. The B&S events at 2 keV_{ee} and 700 eV_{ee} are respectively depicted in Figure 6a&b.

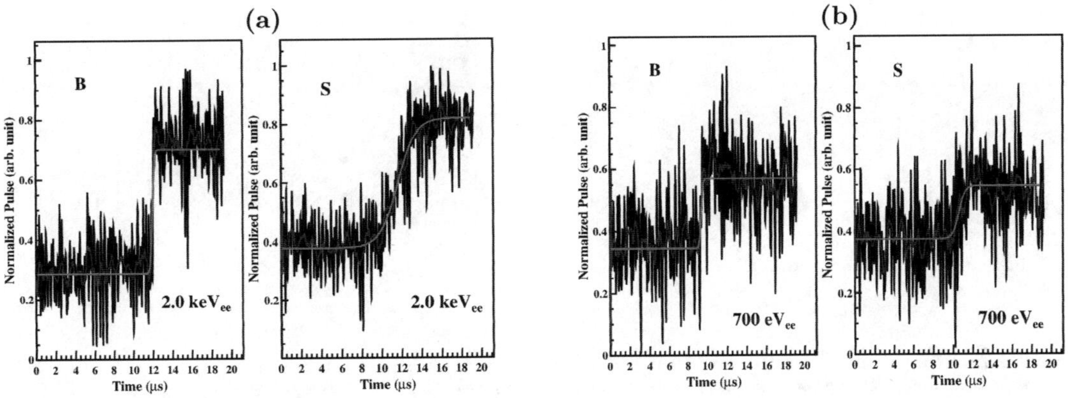

Figure 6. Typical S/B events at (a) 2 $keVee$ and (b) 700 $eVee$ energy with raw pulses shown in black. The smoothed (blue) pulses, together with the best-fit functions(red) are also superimposed.

The rise-time of TA pulse were evaluated by using the hyperbolic tangent function:

$$\frac{A_0}{2}\tanh(\frac{t - t_0}{\tau}) + P_0 \tag{1}$$

where, A_0, P_0 and t_0 are amplitude, pedestal offset and timing offset, respectively.

P_0 and A_0 are evaluated from the TA-pulses through the difference of asymptotic levels, respectively. The time difference as a function of energy between the TA-edge and the DAQ-trigger instant defined by SA signals is predetermined, and provides constraints on t_0.

TA amplitude becomes comparable to noise fluctuations at threshold. Therefore, TA pulses were first smoothed by Savitzky-Golay filters [19] with (τ,t_0) as free parameters. The obtained fitting parameters were then employed as initial values to fit raw TA pulses and final fitting parameters were obtained. The smoothed and final fit are shown by blue and red lines in Figure 6.

The scatter plot of τ versus T for $AC^{-} \otimes CR^{-}$ events is shown in Figure7a&b for pPCGe and nPCGe, respectively. For pPCGe, events with τ less(greater) than a selected cut-value τ_0

(=1.23 μs) are categorized as B(S). In contrast, nPCGe does not exhibit anomalous surface events, as illustrated in Figure 7b where the S-band is totally absent.

The width of measured B and S bands at $T > 1.5\ keV_{ee}$ is much less than separation of bands from τ_0. Therefore, measured τ provides valid information on locations of events and efficiently differentiate S and B events. This behavior also manifests as distinct two band structures in Figure 7a, with a small fraction (about 8% within 3−6 keV_{ee} of AC$^-\otimes$ CR$^-$) of events in intermediate transition zone. By studying the corresponding fractions of events with ^{241}Am (<1%) and ^{137}Cs (7.5%) γ-sources, a thickness of 0.16 mm for this zone was derived. The choice of τ_0 is equivalent to defining the spatial borderline between B/S within this transition thickness. This gives rise to a systematic uncertainty in evaluation of pPCGe fiducial mass. It translates to about 3% of total error at 500 eV_{ee}.

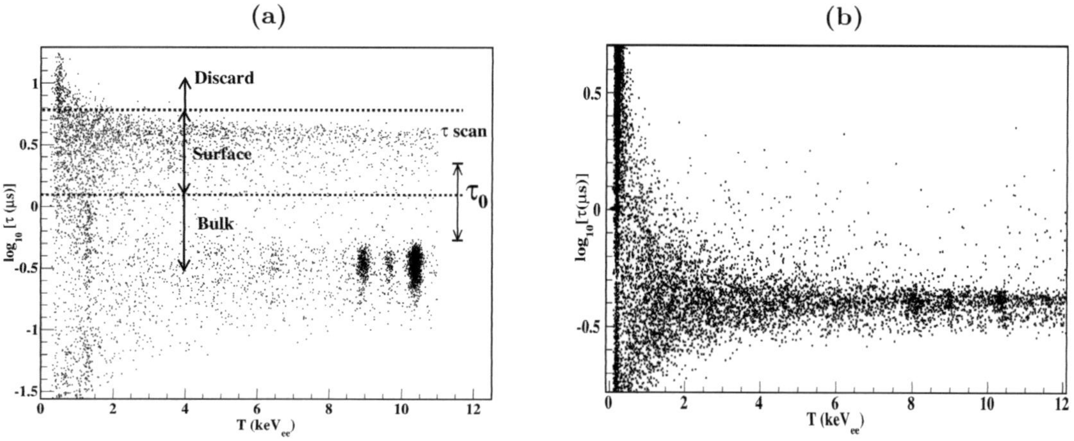

Figure 7. (a) pPCGe τ-distribution at low background configuration. τ_0 defines the surface-bulk selection criteria. (b) nPCGe τ-distribution at low background configuration and demonstrates the absence of anomalous surface events.

The width of the measured B and S band at $T < 1.5\ keV_{ee}$ is comparable to band separation and leads to merging of bands. Therefore, there exists contamination in B and S events. The evaluation of leakage factors and process to correct the measured spectra are further discussed in the section 3.1.

3.1. BS cut efficiencies measurement and correction

The calibration of BS cut requires measurement of bulk signal retaining ($\epsilon_{\rm BS}$) and surface background suppressing ($\lambda_{\rm BS}$) efficiencies. These efficiencies can be obtained by relating the observed rates (B,S) and actual rates (B$_0$,S$_0$) [15].

At lower energy, (B,S) and (B$_0$,S$_0$) are related by the following coupled equations:

$$\begin{aligned} \mathrm{B} &= \epsilon_{\rm BS} \cdot \mathrm{B_0} + (1-\lambda_{\rm BS}) \cdot \mathrm{S_0} \\ \mathrm{S} &= (1-\epsilon_{\rm BS}) \cdot \mathrm{B_0} + \lambda_{\rm BS} \cdot \mathrm{S_0} \ , \end{aligned} \quad (2)$$

with an additional unitarity constrain: B$_0$+S$_0$=B+S. The normalization assignment (B$_0$,S$_0$)=(B,S) is made on events within T_0=2.7-3.7 keV$_{\rm ee}$. It is equivalent to setting $\epsilon_{\rm BS}$ and $\lambda_{\rm BS}$ to 1.0. This energy range is selected because it is above the complications of L-shell X-rays at ~1 keV$_{\rm ee}$ as well as the physics region in dark matter analysis.

The derivation of (ϵ_{BS},λ_{BS}) therefore requires at least two measurements of (B,S) where the actual rates (B_0,S_0) are known. The data with ^{241}Am, ^{137}Cs and *in situ* cosmic-ray induced fast neutrons are used for obtaining ϵ_{BS} and λ_{BS}.

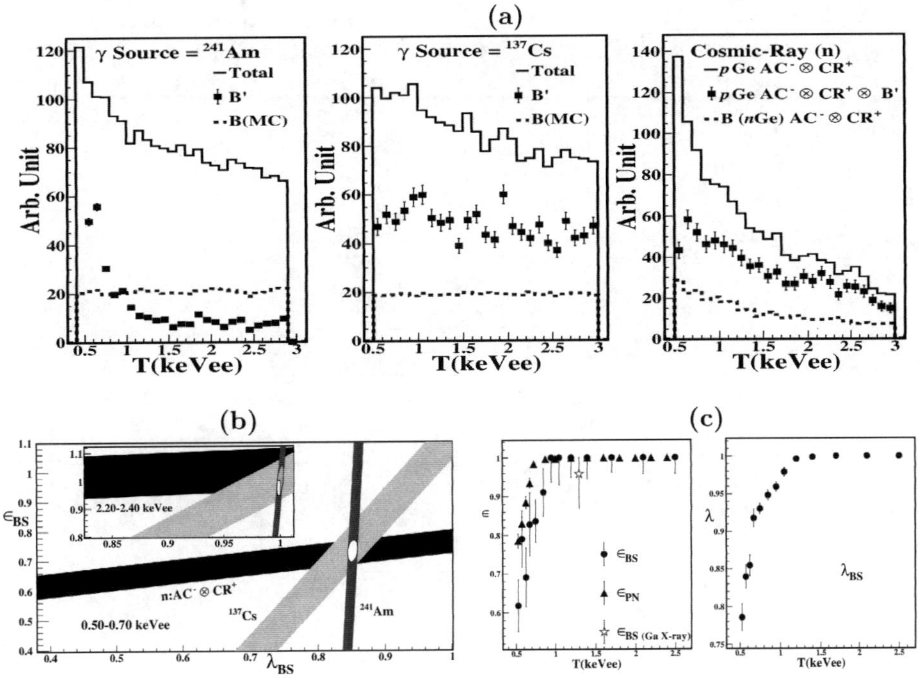

Figure 8. The derivation of ($\epsilon_{BS}, \lambda_{BS}$). (a) The measured Total and B′ spectra from pPCGE with surface-rich γ-ray (^{241}Am, ^{137}Cs) and bulk-rich cosmic-ray induced neutrons. They are compared to reference B-spectra acquired via simulations for γ-rays and nPCGe measurement for cosmic-neutrons. (b) Allowed bands at threshold and at a high energy band. (c) The measured (ϵ_{BS},λ_{BS}) and ϵ_{PN} as functions of energy. Independent measurement on ϵ_{BS} with Ga-L X-rays is included.

The three chosen sources play a complementary role and these data samples are displayed in Figure 8a.

(I): Surface-rich γ-ray sources events– Calibrations with both low and high energy γ-sources (^{241}Am at 60 keVee and ^{137}Cs at 662 keVee, respectively) were performed. As displayed in Figure 8a, the measured B′-spectra are compared to the reference B derived from full simulation with surface layer thickness of 1.16 mm as input. The simulated B-spectra due to external γ-sources over a large range of energy are flat for $T<10$ keVee.

(II): Bulk-rich cosmic-ray induced fast neutrons events – A 523 g first-of-its-kind nPCGe detector was constructed. The components and dimensions are identical to pPCGe. The surface of nPCGe is a p^+ boron implanted electrode of sub-micron thickness. There are no anomalous surface effects. Data were taken under identical shielding configuration. The trigger efficiency was 100% above T=500 eVee, and energy calibration was obtained from standard internal X-ray lines. The $AC^-\otimes CR^+$ condition selects cosmic-ray induced fast neutron events without associated γ-activities, which manifest mostly (~85%) as bulk events. Accordingly, $AC^-\otimes CR^+$ spectrum in nPCGe is taken as B-reference and compared with $AC^-\otimes CR^+$ in pPCGe.

Journal of Physics: Conference Series **606** (2015) 012011 doi:10.1088/1742-6596/606/1/012011

Using calibration data (I) and (II), (ϵ_{BS},λ_{BS}) are measured by solving the coupled Equation 2. Standard error propagation formulae are adopted to derive their uncertainties using errors in (B,B′,S′) as input. As examples, three allowed bands at threshold and at a high energy band are illustrated in Figure 8b. The different orientations of bands are consequences of different depth distributions of the samples, which give rise to different B:S ratios. The bands have common overlap regions, indicating the results are insensitive to event locations. The surface-rich γ-events and bulk-rich cosmic-ray induced neutron-events play complementary roles in constraining λ_{BS} and ϵ_{BS}, respectively. The results are depicted in Figure 8c, with ϵ_{PN} overlaid. By comparing the measured *in situ* Ga-L X-ray peak at 1.3 keVee after BS-selection to that predicted by corresponding K-peak at 10.37 keVee, a consistent ϵ_{BS} is independently measured.

The efficiency-corrected (B_0,S_0) of physics samples can then be derived with the help of measured (ϵ_{BS}, λ_{BS}) by the solution of Equation 2, and is given by

$$\begin{aligned} B_0 &= \frac{\lambda_{BS} \cdot B - (1-\lambda_{BS}) \cdot S}{(\epsilon_{BS} + \lambda_{BS} - 1)} \\ S_0 &= \frac{\epsilon_{BS} \cdot S - (1-\epsilon_{BS}) \cdot B}{(\epsilon_{BS} + \lambda_{BS} - 1)} \quad . \end{aligned} \tag{3}$$

The formulae can be understood as: $B_0(S_0)$ should account for loss of efficiency in the measurement of B(S) in the first positive term, followed by a subtraction of leakage effect from S(B) in second negative term.

3.2. Error sources and assignment

The errors on (ϵ_{BS},λ_{BS}) are shown in Figure 8c. They are derived from global fits on allowed bands in Figure 8b. Standard error propagation techniques were applied to derive the resulting uncertainties on (B_0,S_0) via Equation 3. It can be noted from the Table 2 that the dominant error contribution is due to the statistical uncertainties of the physics data (as observations are associated with low count rate) and is boosted by $\frac{1}{(\epsilon_{BS}+\lambda_{BS}-1)}$.

Table 2. The contribution of uncertainties on $AC^- \otimes CR^- \otimes B_0$ spectrum. † Errors are combined in quadrature and total error is normalized to 1.0.

Energy Bin	$0.50-0.55$ keV$_{ee}$	$1.90-1.95$ keV$_{ee}$
Measurement and Total Error (kg^{-1}keV^{-1}day^{-1})	10.6±5.0	6.1±1.6
Relative Contributions to Total Error† :		
I) Uncertainties on Calibration (ϵ_{BS},λ_{BS}) from Figure 8c :	0.26	<0.03
II) Measurement Error on B_0 from Eq. 3 :		
Statistical Errors of (B,S)	{ 0.41	{ 0.99
Scaling by $1/(\epsilon_{BS}+\lambda_{BS}-1)$	2.29	1.00
Combined	0.95	0.99
III) Systematic Uncertainties due to Parameter Choice :		
(i) Rise-time Cut-Value τ_0	0.12	0.09
(ii) Fiducial Mass from Choice of τ_0	0.03	0.06
(iii) Normalization Range	{ 0.13	{ 0.07
(iv) (B_0,S_0)=(B,S) at Normalization	0.08	0.03
(v) Choice of Discard Region	0.05	0.001
Combined Systematic Error	0.20	0.12

4. Results and discussion

4.1. Energy spectrum

The $AC^- \otimes CR^-$ tagged events from pPCGe data taken at KSNL at various stages of analysis are depicted in Figure 9a. The measured-B and corrected-B_0 spectra are almost identical at $T > 1.5$ keV_{ee}, this is a direct consequence of $\epsilon_{BS} = \lambda_{BS} = 1$. At low energy, efficiency-correcting and background-subtracting effects compensate each other in this data set. After subtracting flat background due to high energy γ-rays and known L-shell X-rays contributions predicted accurately by measured higher energy K-peaks, the residual spectrum is shown in inset of Figure 9b.

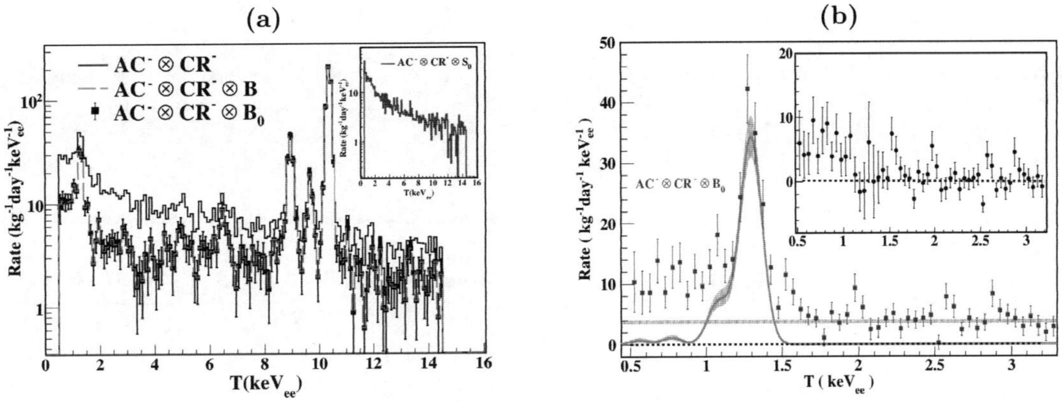

Figure 9. (a) Measured and corrected $AC^- \otimes CR^-$ tag spectra. (b) Shown in magenta are flat background due to high-energy γ-rays from ambient radioactivity, and contributions from the L-shell X-rays. The inset depicts the residual spectrum after background subtraction.

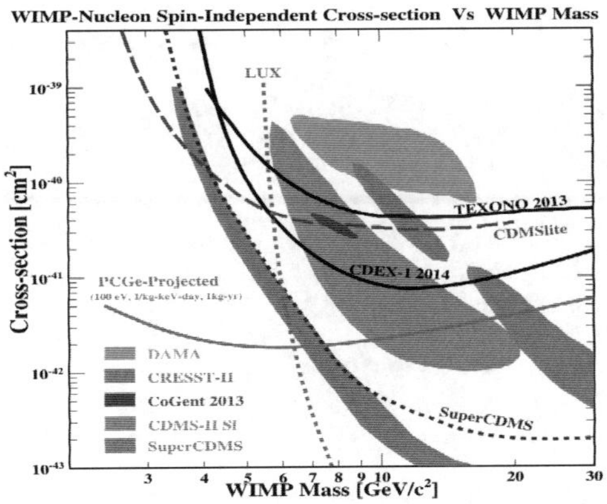

Figure 10. The 90% confidence level upper limit on $\sigma^{SI}_{\chi N}$ from TEXONO, CDEX-1 and other benchmark experiments.

4.2. Constrains on χN spin-independent interaction

The spectrum corresponding to χN candidate events is depicted in inset of Figure 9b. The Constraints on $\sigma^{SI}_{\chi N}$ were then derived via "binned Poisson" method with conventional astrophysical models[8] (local density of 0.3 GeV/cc and Maxwellian velocity distribution with v_0=220 km/s and v_{esc}=544 km/s). The event rates of χN spin-independent interaction cannot be larger than residual spectrum. The Ge quenching function is derived with TRIM software.

Journal of Physics: Conference Series **606** (2015) 012011 doi:10.1088/1742-6596/606/1/012011

Exclusion plots for $\sigma_{\chi N}^{SI}$ versus m_χ at 90% confidence level from TEXONO [11] and CDEX-1 [14] are displayed in Figure 10, with other benchmark results superimposed [14]. The projected sensitivity at 100 eV_{ee} threshold, 1 kg-year data size and background at 1 cpkkd is also shown.

4.3. Constrains on electromagnetic properties of neutrino

The measurable spectra due to weak interactions, neutrino magnetic moments at $\mu_\nu = 10^{-11}$ μ_B and milli-charge ($|\delta_Q|$) at 10^{-12} for reactor neutrino flux of 10^{13} $cm^{-2}s^{-1}$ are depicted in Figure 1. These are compared with various data set [3] and the corresponding limits at 90% CL are listed in Table 3. The limits on μ_ν and $|\delta_Q|$ from nPCGe data are 26.0×10^{-11} μ_B and 2.1×10^{-12} e_0, respectively. μ_B and e_0 are Bohr magneton and electron charge, respectively.

Table 3. A summary of experimental limits at 90% CL on neutrino electro-magnetic parameters using selected reactor neutrino data. The projected sensitivities of measurements at the specified realistically achievable experimental parameters are also shown.

Data Set	Reactor-$\bar{\nu}_e$ Flux ($\times 10^{13}$ $cm^{-2}s^{-1}$)	Data Strength Reactor ON/OFF (kg-days)	Analysis Threshold (keV)	Bounds at 90% CL μ_ν (10^{-11} μ_B)	$|\delta_Q|$ (10^{-12} e_0)
CsI(Tl) - TEXONO [20]	0.64	29882.0/7369.0	3000	< 22.0	< 170
HPGe - GEMMA [21]	2.7	1133.4/280.4	2.8	< 2.9	< 1.1
nPCGe - TEXONO [6]	0.64	124.2/70.3	0.3	< 26.0	< 2.1
Projected [6]	2.7	800/200	0.1	< 1.7	< 0.06

References

[1] Giunti C and Studenikin A 2014 *arXiv:* **1403.6344** and references therein
[2] Wong H T and Li H B 2005 *Mod. Phys. Lett.* **A 20** 1103 and references therein
[3] Li H B 2003 *Phys. Rev. Lett* **90** 131802; Wong H T 2007 *Phys. Rev. D* 75 012001
[4] Davidson S, Gorbahn M and Santamaria A 2005 *Phys. Lett. B* **626** 151; Bell N F 2005 *Phys. Rev. Lett.* **95** 151802; Bell N F 2006 *Phys. Lett. B* **642** 377
[5] Kopeikin V I 1997 *Phys. Atom. Nucl.* **60** 2032; Fayans S A, Mikaelyan L A and Sinev V V 2001 *Phys. Atom. Nucl.* **64** 1475; Kouzakov K A, Studenikin A I and Voloshin M B 2011 *Phys. Rev.D* **83** 113001; Chen J W 2014 *Phys. Lett. B* **731** 159
[6] Chen J W 2014 *Phys. Rev. D* **90** 011301(R)
[7] Yue Q 2004 *HEP and Nucl. Phys.* **28** 877; Wong H T 2006 *J. Phys. Conf. Ser.* **39** 266; Wong H T 2011 *Int. J. Mod. Phys. D* **20** 1463
[8] Drees M and Gerbier G 2014 Chin. Phys. C**38** 353 and references therein
[9] Bertolini G and Coche A 1968 *Semiconductor detectors* Elsevier North Holland Amsterdam
[10] Wong H T 2008 *Mod. Phys. Lett. A* **23** 1431; Lin S T 2009 *Phys. Rev. D* **79** 061101(R)
[11] Li H B 2013 *Phys. Rev. Lett.* **110** 261301
[12] Soma A K 2014 *arXiv:* **1411.4802 [physics.ins-det]** and references therein
[13] Liu S K 2014 *Phys. Rev. D* **90** 032003
[14] Kang K J 2013 *Chin. Phys. C* **37** 126002; Zhao W 2013 *Phys. Rev. D* **88** 052004; Yue Q 2014 *Phys. Rev. D* **90** 091701(R)
[15] Li H B 2014 *Astropart. Phys.* **56** 1
[16] Luke P N 1989 *IEEE Trans. Nucl. Sci. B* **36** 926
[17] Barbeau P S, Collar J I and Tench O 2007 *J. Cosmol. Astropart. Phys. B* **09** 009; Aalseth C E 2008 *Phys. Rev. Lett.* **101** 251301
[18] Aguayo E 2013 *Nucl. Instrum. Meth. A* **701** 176
[19] Press W H 2002 *Numerical Recipes in C* Second edition Cambridge University Press Chapter 14 Page 650
[20] Deniz M 2010 *Phys. Rev. D* **82** 033004; *arXiv[hep-ph]* **1006.1947**
[21] Beda A G 2012 *Adv. High Energy Phys.* **2012** 350150; Beda A G 2013 *Phys. Part. Nucl. Lett.* **10** 139

Journal of Physics: Conference Series **606** (2015) 012012 doi:10.1088/1742-6596/606/1/012012

High purity germanium crystal growth at the University of South Dakota

Guojian Wang[1a], Hao Mei[a], Dongming Mei[a], Yutong Guan[a], Gang Yang[a]

a. Department of Physics, University of South Dakota, Vermillion, SD, 57069

Email: Guojian.Wang@usd.edu

Abstract. High-purity germanium crystal growth is challenging work, requiring the control of individual crystal properties such as the impurity distribution, the dislocation density, and the crystalline structure. Currently, we grow high-purity germanium crystals by the Czochralski method in our laboratory in order to understand the details of the growing process, especially for large diameter crystals. In this paper, we report the progress of detector-grade germanium crystal growth at the University of South Dakota.

1. Introduction

Recently, Germanium-based experiments have reported a possible dark matter signature (CoGeNT)[1] and the claimed discovery of neutrinoless double beta decay[2]. Therefore, the detectors made by high purity germanium (HP-Ge) crystals are preferred for use in future experiments. However, the commercially available germanium crystals will have to be customized to meet the sensitivity requirements set by next generation experiments, which prefer the crystals to be grown in an underground environment that minimizes the cosmogenic activation. It is almost impossible to request commercial companies to meet both production rate and sensitivity for next generation experiments according to our investigation. Thus, additional capacity for growing high-purity germanium crystals with a research entity is a key element in the success of these underground experiments. Motivated by large demands from ultra-low background experiments searching for dark matter and neutrinoless double-beta decay with germanium-based detectors such as Super-CDMS [3], CRESST [4], Majorana [5], GERDA [6], and CDEX [7], we are in cooperation with Sanford Underground Research Facility (SURF) to grow HP-Ge crystal and fabricate it into detectors underground. Without the effect of cosmic rays, our detector will fit the demands for germanium-based neutrinoless double-beta decay and dark matter experiments. We want to create state-of-the-art detectors to advance neutrinoless double-beta decay and dark matter exploration research and technology while simultaneously paving the way for infrastructure to support an underground laboratory for crystal growth, and detector. However, much R&D needs to be done on the surface to grow detector-grade crystals.

There are three challenges in crystal growth. Firstly, the impurity level should be lower than 10^{10} cm^{-3} for the requirement of detector fabrication with depletion region higher than one centimeter. The

[1] Guojian.Wang@usd.edu

impurity concentration in germanium ingots should be purified by Zone Refining method down to 10^{11} cm^{-3}.

Secondly, the HP-Ge crystals must be grown in high purity hydrogen atmosphere to reduce the silicon oxide going into crystal to form trapping centers. Hydrogen has large thermal conductivity, convective heat transfer coefficient and lower viscosity. These make the thermal field difficult to be controlled. Therefore, it is a challenge for growing crystal to control the dislocation density in the range of 100 -10,000 cm^{-2} and make the dislocation uniformly distributed in the crystal.

In the end of crystal growth process, all the germanium melt should be grown out. The thermal expansion coefficient in quartz is much less than germanium. Therefore, even a drop of germanium melt will break the valuable quartz crucible.

Here, we report the current progress of HP-Ge crystal growth in our lab.

2. Experimental

2.1. Crystal growth

The HP-Ge crystals are grown by the Czochralski method [8, 9]. Before each crystal growth, the quartz crucible, the raw Ge rod and the Ge seed were cleaned by sequence of DI water (18MΩ*cm), acetone (A.R), methanol (A.R). Then the quartz crucible was etched by HF : HNO_3 (1:3) solution for 8 minutes and the Ge rod and the Ge seed were etched by HF : HNO_3 (1:3) solution for 4 minutes. Following the etching, the quartz crucible, the raw germanium rods and the germanium seed were rinsed by DI water and blown dry by nitrogen gas. The quartz crucible was located in a graphite crucible, which was used to transfer the high frequency electromagnetic field into heat to melt the germanium rods in the quartz crucible. The crystal growth was carried out in hydrogen atmosphere with flow rate 1.5 l/min. After the germanium rods are melted, the temperature of the melt is cooled down to the melting point (937^{0}C) of germanium. The seed is lowered to touch the melt and the crystal growth process is started. When a germanium crystal is pulled along ⟨1 0 0⟩ direction, four (1 1 1) planes are oblique to the pull axis [10]. Therefore, most of the dislocations with Burgers vectors of (a/2) [1 0 1] type will grow out and terminate at the crystal surface. Dash's neck growth technique [11, 12] is introduced to reduce the dislocations generated in the seed through thermal shock. Then, the diameter is increased as the shoulder of the crystal is grown. The diameter is kept constant in order to grow the body of the crystal. During growth of the tail the diameter is decreased and the last remainder of the melt is used up to avoid breaking the quartz crucible [9]. Figure 1 illustrates the four processes of HP-Ge crystal growth. Figure 2 shows the grown crystals with different diameters: Ø 3.5, Ø 10.5 and Ø 12.7 cm. Currently, we can grow crystals with diameter from 3.5 to 12.7 cm and weight from 0.9 to 6.5 Kg.

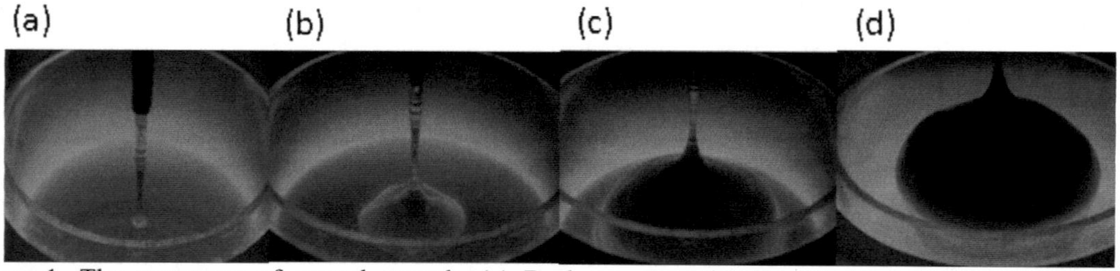

Figure 1. The processes of crystal growth, (a) Dash process, (b) Shouldering process, (c) Equal diameter growth, (d) Ending process.

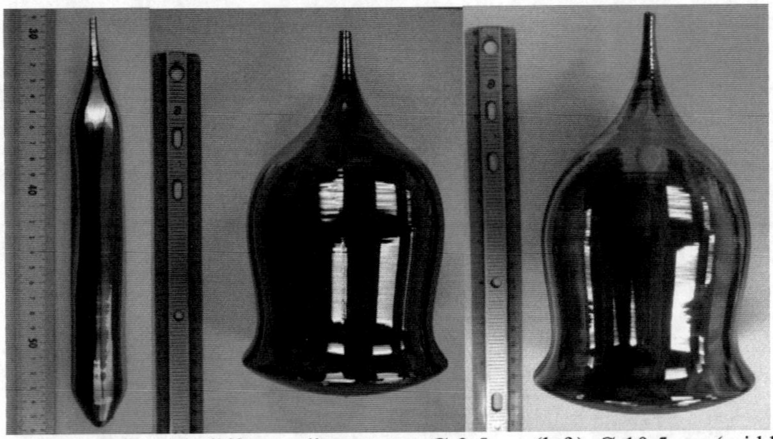

Figure 2.The grown crystals with different diameters : Ø 3.5 cm (left), Ø 10.5 cm (middle) and Ø 12.7 cm (right).

2.2. Characterization of grown crystals
The grown crystals are oriented and studied by X-ray diffraction using a Rigaku Ultima IV X-ray diffractometer. The Olympus BX40 microscope is used to count the dislocation density in the crystals. The net carrier concentration and material type of select samples are determined using Hall effect measurements. The samples are measured at 77 K using the Hall effect with the Van der Pauw geometry. An Ecopia HMS-3000 with a 0.55 T permanent magnet is used to make the measurements. For these measurements, square samples with dimensions of 1.5x1.5x0.16 cm^3 are cut from select wafers.

3. Results and discussion

Characterization of a grown crystal provides necessary feedback information to make improvements in the growth procedure. We use the standard characterization procedures adopted by the early pioneers in the field of germanium crystal growth [13]. Figure 3 shows the X-ray rocking curves from middle part of ϕ3.5, ϕ10.5, and ϕ12.7 cm crystals, respectively. It is observed that there is an extremely strong peak at ω of 29.99° on the XRD patterns of three samples with FWHM 0.035-0.038°. This diffraction peak is the reflection of (400) in germanium indicates that the whole crystal has a very nice <100> orientation. The FWHM of these peaks is from 0.035 to 0.038° confirming the high quality of the crystals.

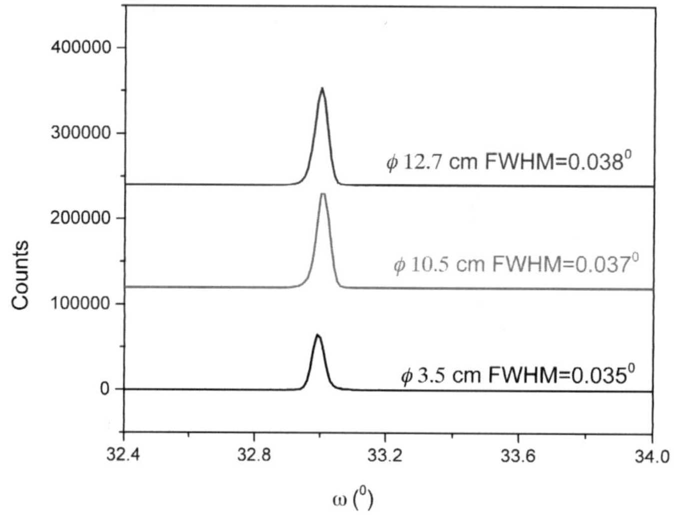

Figure 3. The X ray rocking curves for different dimension crystals.

HP-Ge crystals were grown in Hydrogen atmosphere. Hydrogen will form hydrogen di-vacancy (V_2H) in germanium crystal, which acts as a trap center in HP-Ge detector. Experience has shown that if the dislocation density in the crystal is more than 100 cm^{-2}, the V_2H trap concentration will be too low to influence radiation detector properties [14]. However, a high dislocation density results in too many defect states, which serve as generation/recombination centers or trap states for charge carriers. The defects lead to increased leakage current of the detector and hence result in reduced detector sensitivity. Deep defects also contribute to flicker noise in the detectors. The maximum dislocation density should not be higher than 10,000 cm^{-2}.

It was known that a high impurity concentration results in an excessively high depletion voltage and risk of breakdown. The net impurity concentration for the crystals is required to be controlled around 10^{10} cm^{-3}, depending on the size of the detectors. Figure 4 shows the typical etch pits for the top, middle and bottom slices samples cut from the Ø 3.5 cm crystal. There is no dislocation was found at the top sample. The dislocation density was counted, and found to be around 4,000 cm^{-2} for the middle and bottom samples. Figure 5 presents the dislocation densities are 3,000 and 7,000 cm^{-2} from top and bottom sample, respectively.

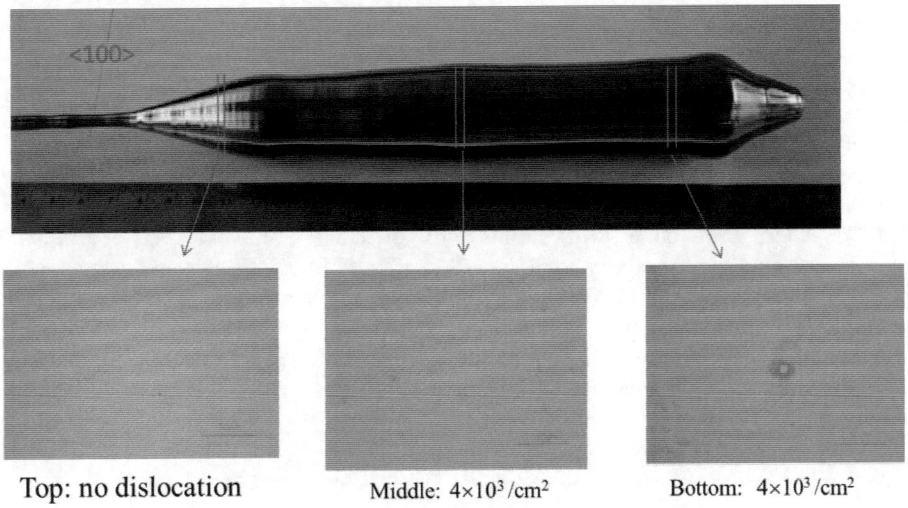

Figure 2. The dislocation densities from top, middle and bottom of Ø 3.5 cm crystal.

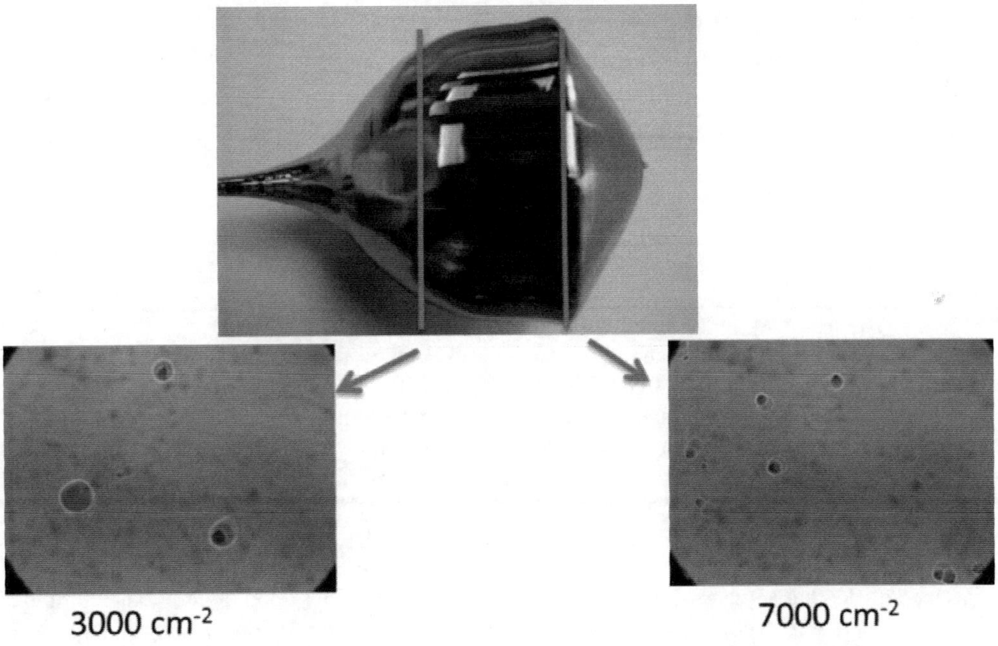

Figure 5. The dislocation densities from top and bottom of Ø12.7 cm crystal

In HP-Ge crystal, the main donors are phosphorus, arsenic, and antimony and the main acceptors are aluminum, boron, and gallium. The net carrier concentration $|N_A-N_D|$ depends on the resistivity (ϱ) of germanium, as following:

$$|N_A-N_D| = 1/(\varrho e \mu) \qquad (1)$$

where μ is mobility at 77K and e is the electron charge (1.6×10^{-19} coulombs).

Hall effect measurements are the effective method to measure such low impurity level (10^{10} cm^{-3}) in HP-Ge. Hall coefficient R_H is measured in terms of the voltage produced across the Hall probes, per unit magnetic field (B) and current (I).

$$R_H = \frac{C}{|N_A - N_D|e} = \frac{\Delta Vt}{IB} \tag{2}$$

where ΔV is the Hall voltage, t the thickness of the plate along the magnetic field. The factor C depends on temperature, crystal type and crystal orientation and is always close to unity [14]. We used Van der pauw geometry for our Hall effect measurements [15]. Table 1 displays the Hall measurements for the grown crystals demonstrate that the impurity levels range from 2.9 – $5.3\text{x}10^{10}/\text{cm}^3$ depending on the grown crystals with different diameter. For p type high purity germanium crystals, the Hall mobility is higher than 40,000 $\text{cm}^2/\text{V*S}$. However, the mobility for n type crystal No.23-14 is around 30,000 $\text{cm}^2/\text{V*S}$. At liquid nitrogen temperature, the lattice scattering was frozen and the impurity scattering is dominant. For hole and electron in germanium crystal, the values of drift mobility are 42,000 and 36,000 $\text{cm}^2/\text{V*S}$ at 77 K [16], respectively. The ratio of Hall mobility to drift mobility is close to 1 at 77K [16]. It confirms the dominant carriers are holes for p type and electrons for n type. For detector fabrication, the Hall mobility of samples should be higher than 25,000 $\text{cm}^2/\text{V*S}$ [13]. The detector fabrication from the grown crystals has been addressed in our lab [17].

Table 1. The Hall Effect measurement results for different diameter crystals.

Crystal number	Net carrier concentration (10^{10}cm^{-3})	Resistivity (Ω*cm)	Mobility ($10^4\text{cm}^2/\text{V*S}$)	Type	Crystal diameter (cm)
30-13	4.433	3184	4.423	P	3.5
20-14	2.917	4536	4.718	P	10.7
23-14	5.321	3892	3.014	N	11.5
24-14	3.673	3786	4.489	P	11.5
35-14	4.160	3158	4.751	P	12.7

The net impurity concentration and the dislocation density for the crystals are required to be controlled within a narrow range of $0.5 - 1.5\times10^{10}\ \text{cm}^{-3}$ and $100 - 10{,}000/\text{cm}^2$, respectively, depending on the size of the detectors. We demonstrate the required impurity as a function of the size of the detector in Figure 6. It was understood that too low of an impurity concentration results in a highly non-uniform electric field; too high of an impurity concentration results in an excessively high depletion voltage and risk of breakdown. Similar issues arise with the dislocation density. It is also demonstrated that dislocation-free crystals do not yield good detectors because dislocations serve as nucleation centers for vacancies and as traps for impurity ions, thus acting as "getters", spatially isolating the defects that serve as traps to charge carriers. On the other hand, a high dislocation density results in too many defect states, which serve as generation/recombination centers or trap states for charge carriers. The defects lead to increased leakage current of the detector and hence result in reduced detector sensitivity. Deep defects also contribute to flicker noise in the detectors.

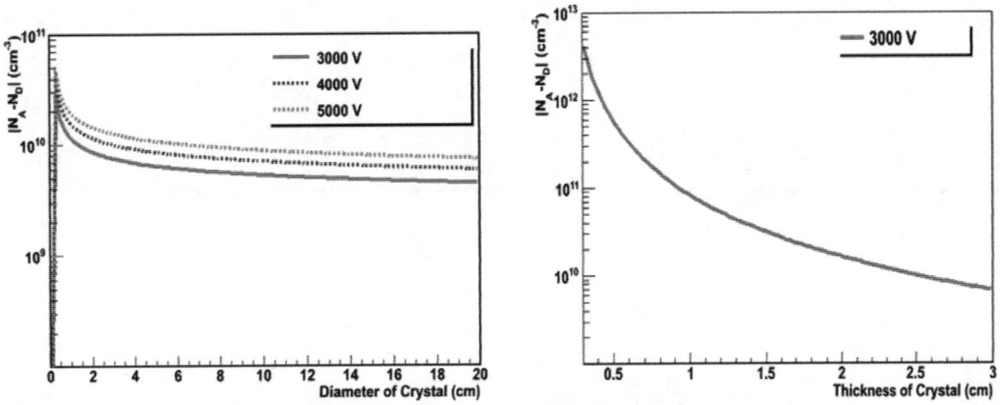

Fig. 6: Left: shown is for a coaxial detector. Right: shown is for a planar detector.

4. Conclusion

We have conducted intensive studies to address challenges in the detector-grade crystal growth and the correlation of these challenges to the use of various methods in the control process. This information allows us to reliably reduce impurity concentrations and dislocation density to acceptable levels on a consistent basis. Consequently, high quality crystals are being grown on a weekly basis at the surface facility located at USD. The grown crystals are in diameters from 3.5 cm to 12.7 cm due to the limitations of the current grower.

Acknowledgement

This work is supported by DOE grant DE-FG02-10ER46709 and the state of South Dakota.

References

[1] Aalseth. C and et al. (CoGeNT Collaboration) 2011 *Phys. Rev.Lett.* **107** 141301
[2] Klapdor-Kleingrothaus H V , DIETZ A, Harney H, and Krivosheina I V 2001 *Mod. Phys. Lett.* A **16** 2409
[3] The CDMS II Collaboration 2010 *Science* **327** 1619
[4] Angloher G and et al.(CRESST Collaboration) 2012 *Eur. Phys. J. C.* **72** 1
[5] Wilkerson J F and et al.(Majorana Collaboration) 2012 *Journal of Physics: Conference Series* **375** 042010
[6] Peiffer P, Motta D, Schoenert S, and Simgen H 2005 *Nucl.Phys.B* **143** 511
[7] Yue Q 2011 *Scientia Sinica Phys, Mech&Astron.* **41** 6
[8] Wang G, Sun Y, Yang G, Xiang W, Guan Y, Mei D, Keller C and Chan Y D 2012 *J. Cryst. Growth* 352 (1), 27 (2012).
[9] G. Wang, Y. Guan, H. Mei, D. Mei, G. Yang, J. Govani, and M. Khizar, *J. Cryst. Growth* **393** 54
[10] O`Mara W C, Herring R B and Hunt L P 1990 *Handbook of Semiconductor Silicon Technology* (New Jersey:Noyes).
[11] Dash W C 1959 *J. Appl. Phys.* **30** 459
[12] Dash W C 1960 *J. Appl. Phys.* **31** 736
[13] 1993 *IEEE Std* **1160-1993** 36
[14] Haller E E, Hansen W L and Goulding F S 1981 *Adv.Phys.* **30**, 45
[15] Van der pauw L J 1958 *Philips Res. Rep.* **13** 1
[16] Morin F and Maita J 1954 *Phys. Rev.* **94** 1525

Journal of Physics: Conference Series **606** (2015) 012012 doi:10.1088/1742-6596/606/1/012012

[17] Wang G, Amman M, Mei H, Mei D, Irmscher K, Guan Y, Yang G, (unpublished)

Journal of Physics: Conference Series **606** (2015) 012013 doi:10.1088/1742-6596/606/1/012013

Study on the Properties of High Purity Germanium Crystals

G Yang [1], H Mei[1], YT Guan, GJ Wang[1], DM Mei[1], K Irmscher[2]

1. Department of Physics, The University of South Dakota, Vermillion, SD 57069, USA
2. Leibniz-Institut für Kristallzüchtung (IKZ), Berlin, Germany

Email: gang.yang@usd.edu

Abstract In the crystal growth lab of South Dakota University, we are growing high purity germanium (HPGe) crystals and using the grown crystals to make radiation detectors. As the detector grade HPGe crystals, they have to meet two critical requirements: an impurity level of ~10^9 to 10^{10} atoms /cm^3 and a dislocation density in the range of ~10^2 to 10^4/ cm^2. In the present work, we have used the following four characterization techniques to investigate the properties of the grown crystals. First of all, an x-ray diffraction method was used to determine crystal orientation. Secondly, the van der Pauw Hall effect measurement was used to measure the electrical properties. Thirdly, a photo-thermal ionization spectroscopy (PTIS) was used to identify what the impurity atoms are in the crystal. Lastly, an optical microscope observation was used to measure dislocation density in the crystal. All of these characterization techniques have provided great helps to our crystal activities.

1. Introduction

In order to meet high-resolution requirements of radiation detectors, high purity germanium (HPGe) single crystals must have a net impurity level of ~10^{10}/cm^3 and a dislocation density between ~10^2 and ~10^4 /cm^2 throughout the entire crystal, which are two very challenging tasks [1-5].

In our lab, we are carrying out to grow this kind of HPGe crystals and using them to make radiation detectors. The detailed process is shown in Fig.1. Firstly, we used zone refining method to purify the raw germanium materials and then measured their electrical properties using van der Pauw Hall measurement system. If the impurity level of zone-refined ingots reaches 10^{10-11}/cm^3, decreased by a factor of 3 from 10^{13-14}/$cm^3$of raw materials, then the purified ingots could be used to grow HPGe crystals. After a crystal was grown, the x-ray diffraction analysis, van der Pauw Hall measurement, PTIS and measurement of dislocation will be used to characterize grown crystals. Once as-grown crystals with an expected orientation have an impurity level 10^{9-10}/cm^3 and dislocation density in the range of 10^{13-14}/c m^3, we will use them to fabricate the detector. Therefore, the characterization of grown crystals is an indispensable step during HPGe crystal growth.

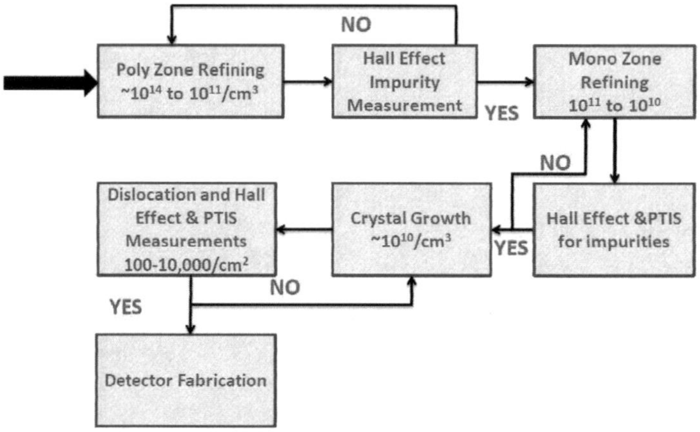

Fig.1 Flow chart of HPGe crystal growth and detector fabrication

2. XRD analysis

XRD analysis is a common and effective method to investigate the crystallographic orientation and structure of the crystals. Contrary to conventional powder measurements, where intensity is monitored in dependence of the diffraction angle 2θ, single crystals should be measured by means of rocking curves. The full width at hall maximum (FWHM) of the peak can be used to characterize crystal quality. Since XRD results can provide the evidence about the quality of grown crystals, they can be used to improve crystal growth process.Fig.2 (a) is an x-ray crystal orientation instrument (XY-3, Dandong, China), which can be used to determine the crystal orientation, such as <100>- or <111>-oriented crystal. After the orientation of the grown crystal was determined, we cut crystal into slide on the diamond wire saw. Furthermore, XRD measurement were carried out to analyzed the crystal quality in the X-ray diffractometer (Rigaku Ultima IV, Japan.) (Fig.2 (b)) with monochromatic CuKα1 radiation (wavelength: λ=1.5406Å), working voltage of 40kV and working current of 44mA.

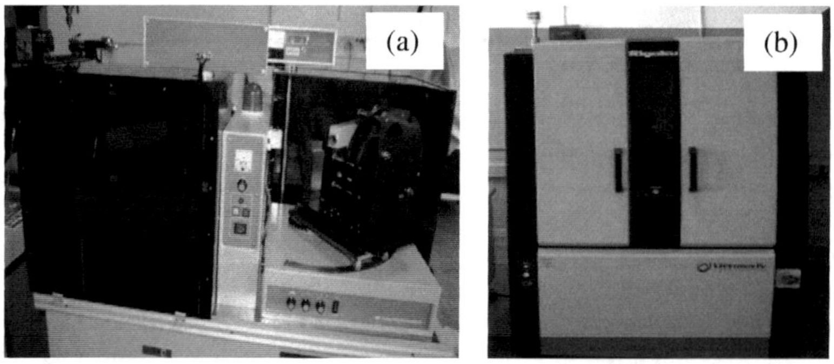

Fig.2 X-ray crystal orientation instrument (a) and x-ray diffractometer (b) at USD

XRD θ/2θ scanning result of slides cut from the <100>-oriented crystal is shown in Fig.3. It is observed that there is an extremely strong (400) peak at 2θ of 66.15° on the XRD pattern of slide sample. Presence of 400 reflection indicates that the whole crystal has a very nice <100> orientation.

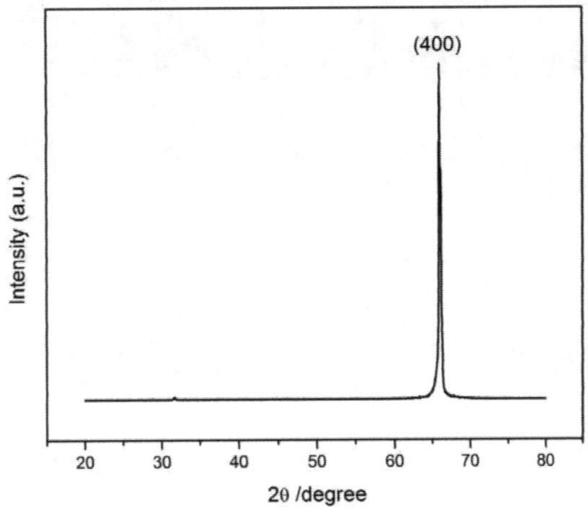

Fig. 3 XRD scanning pattern of <100>-oriented crystal of sample from the head of crystal

3. van der Pauw Hall measurement

van der Pauw Hall measurement is a standard measurement technique to characterize mechanical properties, such as carrier concentration, mobility and resistivity of semiconductors. From measured results, we can know the impurity level of semiconductors. In our lab, we used van der Pauw Hall method to measure the electrical properties of zone-refined ingots and crystals, which combines van der Puaw technique [6] for resistance measurement of thin sample and Hall effect technique [7] for mobility measurement of sample. Usually, we cut three slides from zone-refined one ingot and two slices from single crystals using diamond wire saw. The square samples with dimension of 15mm×1.5mm×1mm were prepared from slices, then polished and etched in $1HF+3HNO_3$ solution, finally dried by nitrogen gas. The measurement was carried out at 77K on the HMS-3000 Hall Effect Measurement System (Ecopia, Korea.), shown in Fig.4. In Hall effect measurement, mobility characterizes the movement ability of charge carriers through a semiconductor under an electric field. At certain temperature, the movement of charge carriers will be scattered by ion impurity atoms, neutral impurity atoms and lattice. Therefore, total mobility is a combination of the mobility from all scattering processes.

Since the lattice will be frozen at 77K, the last two terms in the above equation have little influence on the total mobility. Therefore, the dominant factors should be scattering from neutral and ionized impurity atoms. For HPGe crystal, the mobility should be more than 26000 cm^2/Vs [7].

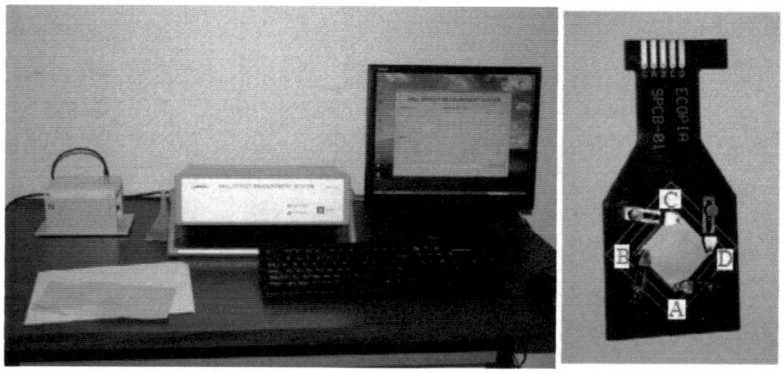

Fig. 4 XRD θ/2θ scanning pattern of <100>-oriented crystal at USD

Table 1 Measured electrical properties of recent zone-refined ingot and crystal

Sample	Carrier Concentration /cm^3)	Mobility(cm^2/V·s)	Resistivity(Ω·cm)
Zone-refined Ingot			
ZR_S1	4.44×10^{11}	4.10×10^4	3.43×10^2
ZR_S2	2.08×10^{11}	4.67×10^4	6.41×10^2
ZR_S3	1.68×10^{11}	4.74×10^4	7.86×10^2
Single Crystal			
Crystal_S1	6.20×10^{10}	4.09×10^4	2.45×10^3
Crystal_S1	4.16×10^{10}	4.75×10^4	3.15×10^3

Table 1show the electrical properties of recent zone-refined ingot and crystal, which involves three samples cut from zone-refined ingot and two samples cut from the single crystal. ZR_S1, _S2 and_S3 represent samples cut from head, middle and tail parts, respectively. Crystal_S1 and _S2 represent samples cut from head and tail parts of crystal, respectively. It is shown that the entire zone-refined ingot has reached the level less than 5 x10^{11}, while the crystal has reached the level of x 10^{10}.

3. PTIS

As a unique technique for identification of residual impurities in high purity semiconductors, the photo-thermal ionization spectroscopy (PTIS) was discovered by T.M. Lifshitz and F. Ya Nad [8-10] when they studied photoconductivity of shallow hydrogenic impurities in germanium. The photoconductivity peaks in PTIS are produced by two-step excitation process. Initially, the electron (or hole) is ionized from the ground state to an excited state by absorption of a photon, and then further excited by a phonon into the conduction band to produce photoconductivity signal [11-12]. Fig.5 shows the two excitation steps of PTIS and setup in the lab.

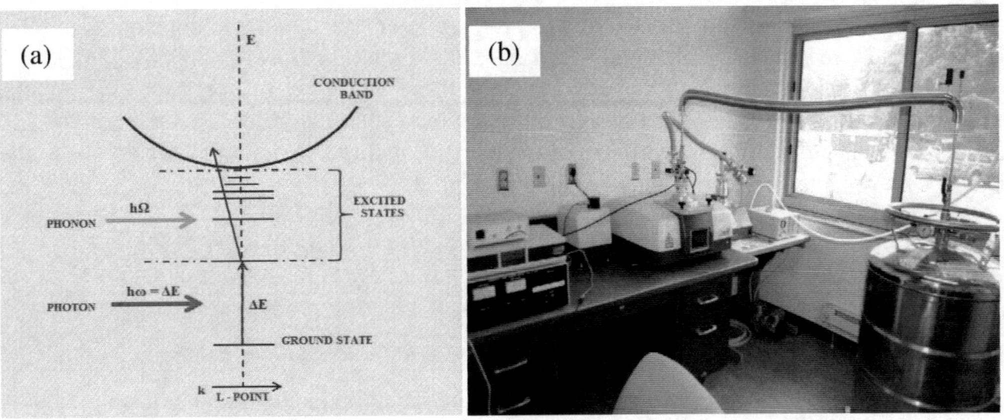

Fig.5 Two excitation of PTIS (a) [] and setup at USD (b)

For PTIS samples, we have prepared two samples with dimensions of 7mm x 7mm x 3mm. Both were cut from the head section of the grown crystals. Before making contacts, the samples were etched for 3 minutes in 3 HNO_3: 1HF solution. Since our PTIS equipment is not ready to use, so these samples were sent to the Leibniz Institute for Crystal Growth (IKZ), Berlin, Germany for the PTIS measurements. The samples were immersed in liquid helium and PTIS spectra were obtained at 7K. The results are shown in Fig. 6.

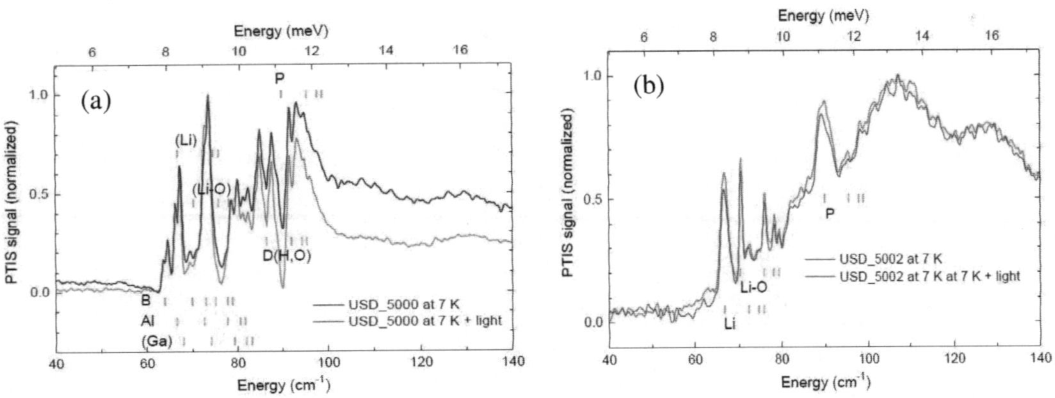

Fig. 6 PTIS spectra of p-type (a) and n-type (b) of two different germanium crystals

Fig.6(a) shows that the dominant impurities are boron and gallium, so the crystal is p-type semiconductor. The dominant impurity atom is phosphor in Fig.6(b), which indicates that the crystal is n-type semiconductor.

3. Dislocation observation

Journal of Physics: Conference Series **606** (2015) 012013 doi:10.1088/1742-6596/606/1/012013

As stated above, dislocation density in detector grade crystals must be controlled in the range of ~10^2 and ~10^4 /cm^2 throughout the entire crystal. Since occurrence of dislocations is directly related to distribution of residual stress in the crystals, which in turn is related to the thermal field in crystal growth, the dislocation density results can be used to improve the thermal field. The sample for dislocation observation was cut from the head part of the crystal. Prior to dislocation observation, the sample surface was polished to mirror-like surface, and then etched in a mixed solution of HF, HNO_3, and CH_3COOH. The dislocation was observed under Nikon Eclipse LV150L microscope (Fig.7(a)). Fig.7(b) shows the etched pits on one site on the slide sample, indicating a dislocation density of 2000/cm^2. The dislocation density on the entire slide is about 1.9×10^3/cm^2, shown in Fig.7(c).

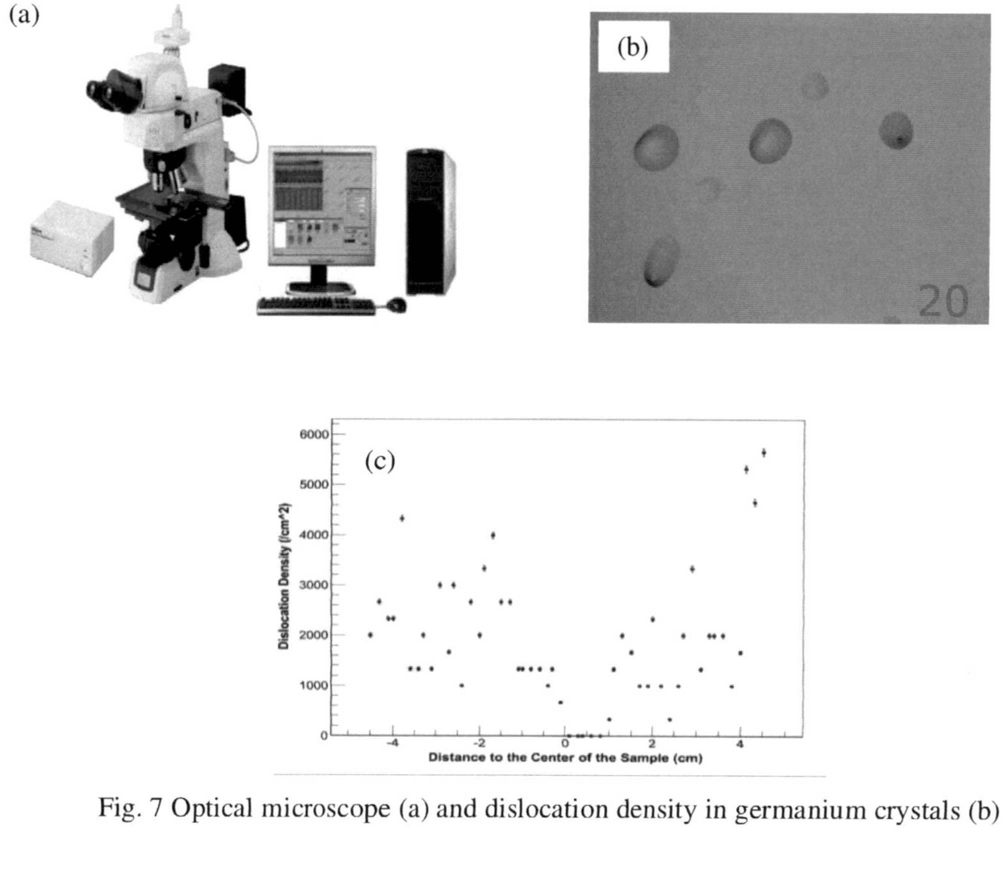

Fig. 7 Optical microscope (a) and dislocation density in germanium crystals (b)

4. Conclusions

In order to assure that the HPGe crystals we grow are detector grade crystals, we have used four kinds of techniques to characterize the grown crystals, such as XRD analysis, van der Pauw Hall effect measurement, PTIS and dislocation observation. XRD technique can be used to grow crystals with the required orientation and high quality. Van der Pauw Hall measurement and PTIS can help us to improve the purification level of the grown crystals. Dislocation results can help us to optimize the thermal field in crystal growth process. With the help of those characterization methods, the grown crystals at USD have an impurity level of $\times 10^{10}/cm^3$ and an average dislocation density of about $2\times 10^3/cm^2$, which meet the requirement for making radiation detectors.

Acknowledgements

This work was supported by supported by DOE grant DE-FG02-10ER46709 and the state of South Dakota.

References:

[1] Claeys C and Simoen E, Germanium-based Technologies from Materials to Devices, Elsevier BV, 2007.
[2] Hansen WL and Haller EE, IEEE Trans.Nucl.Sci. 21(1973)251-259.
[3] Haller EE et al, Advances in Physics 30(1981) 93-138.
[4] Haller EE et al, IEEE Trans.Nucl.Sci. NS-23(1976)81-87.
[5] Hansen WL and Haller EE, IEEE Trans.Nucl.Sci. NS-21(1973) 251- 259.
[6] Van der Pauw LJ, Philips Research Report 13, 1(1958).
[7] Schroder DK 2006 Semiconductor Materials and Devices Characterization, (A John Wiley & Sons, InC., Publication).
[8] Lifschitz TM and Nad FYa, Sov. Phys. Doklady, 10, No. 6, 532(1965).
[9] Sidorov VI and Lifschitz TM, Sov. Phys. Solid State, 8, No. 8, 2000 (1967).
[10] Kogan SHM and Sedunov, Sov. Phys. Solid State, 8, No. 8, 1898 (1967).
[11] Haller EE et al, IEEE Trans.Nucl.Sci. NS-22(1975) 127- 134.
[12] Darken LS, J. Appl. Phys., 65(3) (1989): 1118-1125.

2nd Workshop on Germanium Detectors and Technologies IOP Publishing
Journal of Physics: Conference Series **606** (2015) 012014 doi:10.1088/1742-6596/606/1/012014

Zone Refinement of Germanium Crystals

G Yang[1], YT Guan[1], FY Jian[1], MD Wagner[1], H Mei[1], GJ Wang[1], SM Howard[2], DM Mei[1], A Nelson[3], J Marshall[4], K Fitzgerald[5], C Tenzin[6], X Ma[2],

1. Department of Physics, The University of South Dakota, Vermillion, SD 57069, USA
2. Department of Materials and Metallurgical Engineering, South Dakota School of Mines and Technology, Rapid City, SD 57701
3. Metallurgical Engineer, North American Stainless, Ghent, KY formerly Research Assistant and MS Candidate, Department of Materials and Metallurgical Engineering, South Dakota School of Mines and Technology, Rapid City, SD 57701.
4. Mechanical Engineer, Applied Controls Equipment, Denver, CO formerly Research Assistant and BS Candidate, Department of Materials and Metallurgical Engineering, South Dakota School of Mines and Technology, Rapid City, SD 57701.
5. Metallurgical Engineer, Logan Aluminum, Lewisburg, KY formerly Research Assistant and MS Candidate, Department of Materials and Metallurgical Engineering, South Dakota School of Mines and Technology, Rapid City, SD 57701.
6. Technical Consultant, Wenatchee, WA formerly Research Assistant and MS Candidate, Department of Materials and Metallurgical Engineering, South Dakota School of Mines and Technology, Rapid City, SD 57701.

Email: gang.yang@usd.edu

Abstract Purification of commercial germanium with an impurity level of ~$10^{13\text{-}14}$ cm^{-3} was successfully conducted in two-step zone refining process under an undiluted high-purity hydrogen gas atmosphere. Results for the first step conducted in graphite boats yielded ingots with an impurity level of ~10^{12} cm^{-3} near the center of the 60 cm long ingots. These center portions were collected and subsequently zone refined in a high purity quartz boat to reach a purity level of ~10^{11} cm^{-3}. The best material achieved in a one step process employing a carbon-coated quartz boat yielded material of purity $8x10^{11}$ cm^{-3}.

1. Introduction

Dark matter and neutrino physics are currently topics of great interest in modern physics. Unknown are such properties as the neutrino's absolute mass and magnetic moment as well as the Dirac or nature of the particle. The Neutrino Science Advisory Group and the Dark Matter Science Advisory Group encourage investigation into the better understanding of both dark matter and neutrinos. The MAJORANA and GERDA experiments plan to use high-purity germanium detectors to investigate neutrino properties [1]. They intend to base their work on the neutrinoless double-beta decay event to determine neutrino properties. Very high resolution and very low noise is required for double-beta decay searches using Ge detectors, because of the low event rate of double-beta decay. A high-purity germanium detector is one of the best candidates for such a detector. To satisfy the high resolution requirements, a germanium (HPGe) single crystal detector must have a net impurity level of ~10^{10} cm^{-3} and a dislocation density between 10^2 cm^{-2} and 10^4 cm^{-2} throughout the entire crystal. Production of such crystals is a very challenging task as described by Hansen and Haller [2]. The zone refinement of commercial germanium ingots to ~10^{11} cm^{-3} level is an essential precursory step to produce detector grade germanium crystals.

The zone refining technique was developed at Bell Telephone Laboratories in the early 1950s. It is a fractional solidification technique based on the separation of impurities through crystallization of a liquid without adding a solvent. Impurities have segregation coefficients across the liquid-solid interface that can differ by an order of magnitude, or more. A long germanium ingot is refined by repeatedly moving an inductively-melted liquid zone from one end of the ingot (head) to the other (tail). After multiple zone-refining passes, the impurities will accumulate at the ends of the ingot depending on whether the segregation coefficient is less than or greater than unity; thus, the middle part of the ingot becomes increasingly purified [3-4].

2. Zone-refining equipment

The two commercial zone refining units shown in Figure 1 were acquired and used to refine germanium at the University of South Dakota (USD). They are designated A (ZRA) and B (ZRB). They were able to handle ingots of 4 kg and 16 kg, respectively. Both have two tubes to simultaneously accommodate two ingots. Each tube has single induction coil for heating. The heating power is supplied by a 200 kHz induction generator [5]. A third zone refiner unit shown in Figure 2 was used for zone refining at the South Dakota School of Mines and Technology (SDSM&T). This zone refining unit was designed and built at SDSM&T.

3. Experimental procedure

Commercially pure germanium ingots with a purity level of 99.999% and specific resistance of 47 Ω·cm were purchased for subsequent zone refining purification. The electrical properties of the starting material were measured at room temperature and 77 K. The resistivity of the commercial ingot was approximately 50 Ω·cm at room temperature and the impurity level was

Journal of Physics: Conference Series **606** (2015) 012014 doi:10.1088/1742-6596/606/1/012014

measured to be (10^{13} to 10^{14}) cm^{-3} at 77 K. The main impurities existing in the starting germanium were identified by photothermal ionization spectroscopy (PTIS) to be boron, phosphorous, and aluminum [5].

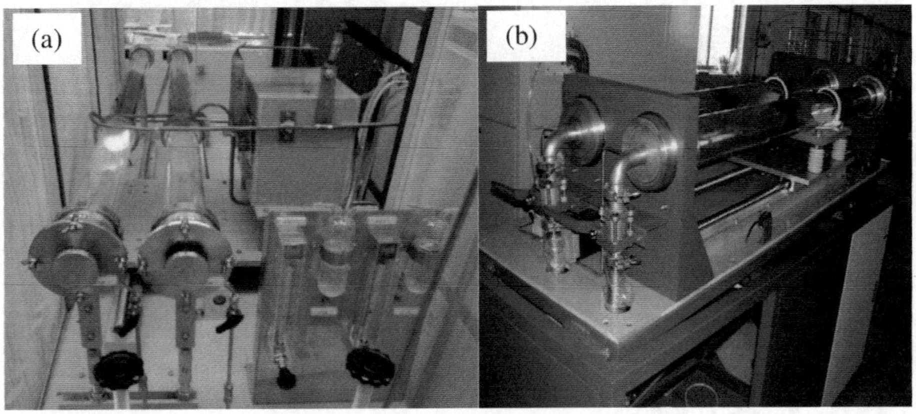

Fig. 1 USD zone refining units a) A and b) B.

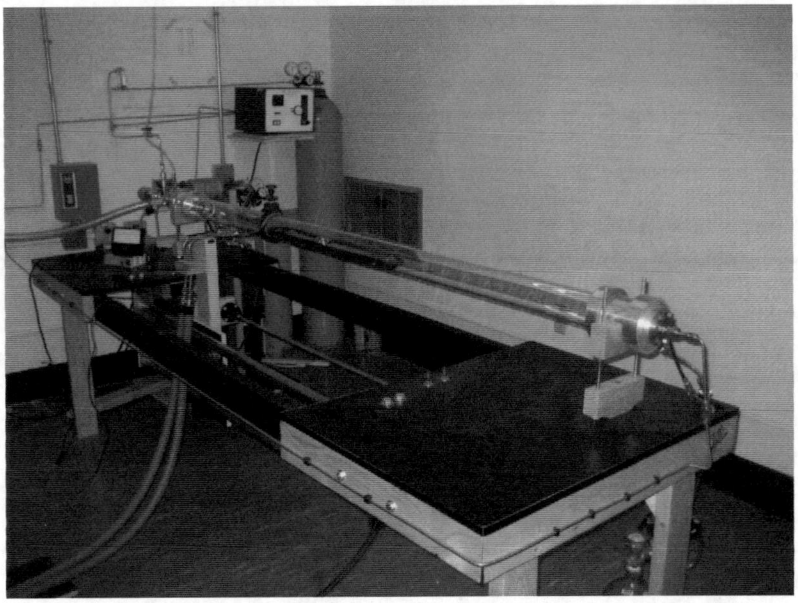

Fig. 2 SDSM&T zone refining unit.

Refractory boats used to hold the germanium during zone refining were made of either very high-purity graphite or ultra-high purity quartz formed from the combustion of high purified silane. Both types of boats were obtained from commercial vendors. Both types of boats were used at USD while only quartz boats were used at SDSM&T. The ingots processed by USD A and SDSM&T zone refining units were 60 cm long with a cross sections of (6 to 12) cm^2 depending on the mass charge (2 to 4) kg. The ingots from USD Unit B were 60 cm long with a cross section of (20 to 40) cm^2.

All germanium, tubes, and boats used in zone refining were very strictly cleaned [5]. Outer quartz tubes were washed using a hydrogen peroxide solution, then washed using distilled deionized water. Graphite boats were cleaned using polyethylene cleanroom wipes. Germanium and quartz boats were etched in a 1 HF: 3 HNO_3 mixture solution to remove any contamination on the surface, then rinsed with distilled and deionized water multiple times, and dried by nitrogen gas. Finally, the quartz boats were coated using high-purity ethylene (99.999%) in a Class-100 clean room.

Once the boat was charged with germanium and secured in the zone refining unit, the enclosed volume of the zone refining unit was evacuated and sealed to verify that there were no leaks. At USD the pumps consisted of high-speed turbo pumps, which typically achieved 1×10^{-5} torr. At SDSM&T the pump was a mechanical pump capable of achieving 2×10^{-4} torr fitted with a molecular sieve trap to prevent oil vapor back streaming. Multiple evacuation and high-purity argon backfill cycles assured adequate removal of oxygen and water vapor from the SDSM&T system before commencing zone refining. The zone refining melt pool for the two 4 kg units was typically (2 to 3) cm wide and was driven the length of the ingot over an eight hour period. The melt pool width was somewhat larger on the 16 kg refining unit and the time somewhat longer so as to keep the velocity of the zone refining the same as for the 4 kg units. The atmosphere above the germanium was ultra-high purity hydrogen. At USD the hydrogen was produced by electrolysis followed by cryogenic condensation of residual water vapor and dryers while at SDSM&T the hydrogen was supplied from ultra-high-purity cylinders. In both locations, the hydrogen passed through a final heated Pd-foil purifier immediately before entering the zone refining units.

The electrical properties such as carrier concentration, mobility and resistivity of ingots were measured at 77 K using van der Pauw Hall measurement [6] system (HMS-3000, Korea). Samples for the Hall measurement were taken from the ingot head, middle and tail; made into square samples; etched in a 1 HF: 3 HNO_3 solution; washed using distilled and deionized water; and dried by nitrogen gas.

4. Results and discussions

The USD zone refining strategy was to initially zone refine in the graphite boat followed by further zone refining in the high-purity quartz boat. The main purpose of zone refining in graphite boat was to remove aluminum. The graphite boat created a reducing environment, which according to Hubbard, Haller, and Hansen [7] promotes the removal of aluminum under what they called Region III conditions: low silicon and low oxygen. Without this step, aluminum does not segregate well in carbon-coated quartz boats (Hubbard, et al Region II: moderate oxygen and moderate silicon). Subsequent zone refining in a quartz boat under hydrogen atmosphere removes boron probably by the formation of silicon-oxygen-boron complexes and phosphorous (Region I: low oxygen and high silicon). Figure 3 shows the photographs of zone-refined ingots in graphite and quartz boats.

After zone refining in the graphite boat, the impurity level of the ingot was ~10^{12} cm^{-3}, which was an improvement by factor of about 100 compared to the raw materials. The resulting material was p-type material as shown in Fig. 4a. After zone refined in quartz boat, the impurity level of all of samples reaches ~10^{11} cm^{-3} as shown in Fig. 4b, which was a further improvement by a factor of about 10 compared to the zone-refined ingots in the graphite boat alone. Because boron and aluminum are acceptor impurities, while phosphorous is a donor impurity, p-type material means that the dominant impurities after zone refining were boron and aluminum.

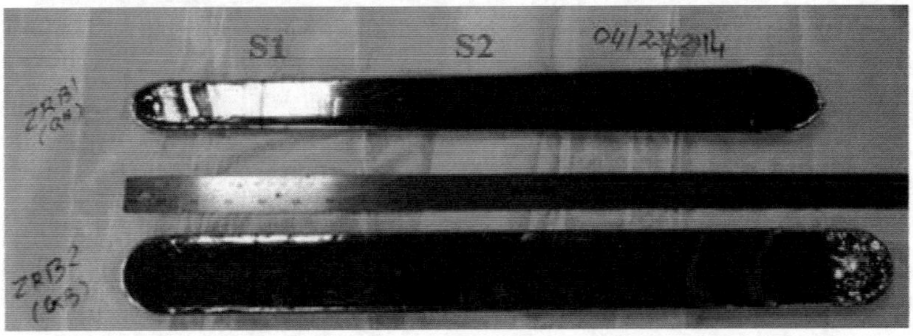

Fig. 3 Zone-refined ingots from quartz (top) and graphite (bottom) boats.

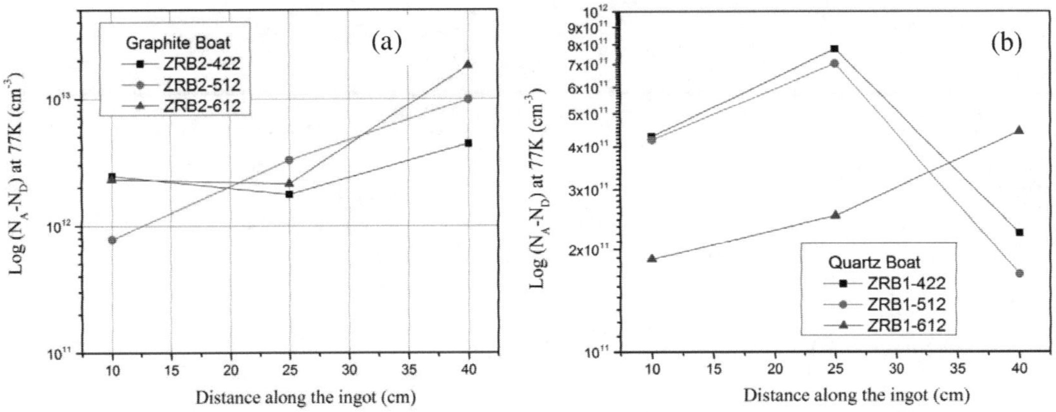

Fig. 4 Net carrier concentration as a function of distance along the ingots at 77 K, from a) graphite boat and b) quartz boat.

Zone refining at SDSM&T was conducted only in high-purity quartz boats coated with carbon under an ultra-pure and undiluted hydrogen atmosphere. Initially carbon was placed with a paraffin flame subsequently replaced with commercial butane and finally very high-purity (>99.9995%) ethylene. The initial carbon coating resulted in ingots ranging from no improvement to levels reaching the ($2x10^{12}$ to $6x10^{12}$) cm^{-3} impurities. As the purity of the carbon coating source was improved, purity levels generally improved finally reaching a result of $8x10^{11}$ cm^{-3} with the high-purity ethylene. In some runs, an attempt was made to purify the carbon coating by holding the coated boat at 1273 K in a high-purity argon-chlorine mixture for several hours followed by several additional hours in only flowing high-purity argon. The results of such attempted purification were no better than those obtained using ultra-pure ethylene for carbon coating.

5. Conclusions

Commercial germanium ingot ($\sim10^{13}$ to $\sim10^{14}$) cm^{-3} may be zone refined in a two-step process: graphite boat followed by high-purity quartz boat. The first step removed and segregated aluminum reaching $\sim10^{12}$ cm^{-3} impurities while the second step segregated phosphorous and boron achieving $\sim10^{11}$ cm^{-3} impurities as determined by the van der Pauw Hall effect measurements at 77 K. Single-step germanium purification in a process involving a quartz boat coated with very high-purity carbon under ultra-high purity hydrogen proved less successful than the two-step process reaching impurity levels of $8x10^{11}$ cm^{-3} and having p-type impurities.
Zone refined germanium was produced that was suitable for growing HPGe single crystals for detectors.

Acknowledgements

This work was supported by supported by DOE grant DE-FG02-10ER46709 and the State of South Dakota.

References

[1] Claeys C and Simoen E, Germanium-based Technologies from Materials to Devices, Elsevier BV, 2007.
[2] Hansen WL and Haller EE, IEEE Trans. Nucl. Sci. 21(1973)251-259.
[3] Haller EE et al, Advances in Physics 30(1981)93-138.
[4] Haller EE et al, IEEE Trans. Nucl. Sci. NS-23(1976)81-87.
[5] Yang G et al, Cryst. Res. Technol., (2014) 269-275.
[6] Van der Pauw LJ, Philips Research Report 13, 1(1958).
[7] Hubbard GS, Haller EE, and Hansen WL, IEEE Trans. Nucl. Sci. NS-25, No. 1(1978)362-70.

The Sanford Underground Research Facility at Homestake

J Heise

Sanford Underground Research Facility, 630 East Summit Street, Lead, SD 57754

E-mail: jaret@sanfordlab.org

Abstract. The former Homestake gold mine in Lead, South Dakota has been transformed into a dedicated facility to pursue underground research in rare-process physics, as well as offering research opportunities in other disciplines such as biology, geology and engineering. A key component of the Sanford Underground Research Facility (SURF) is the Davis Campus, which is in operation at the 4850-foot level (4300 m.w.e.) and currently hosts two main physics projects: the LUX dark matter experiment and the MAJORANA DEMONSTRATOR neutrinoless double-beta decay experiment. In addition, two low-background counters currently operate at the Davis Campus in support of current and future experiments. Expansion of the underground laboratory space is underway at the 4850L Ross Campus in order to maintain and enhance low-background assay capabilities as well as to host a unique nuclear astrophysics accelerator facility. Plans to accommodate other future experiments at SURF are also underway and include the next generation of direct-search dark matter experiments and the Fermilab-led international long-baseline neutrino program. Planning to understand the infrastructure developments necessary to accommodate these future projects is well advanced and in some cases have already started. SURF is a dedicated research facility with significant expansion capability.

1. Introduction

Many disciplines benefit from access to an underground facility dedicated to scientific research, including physics, biology, geology, engineering, and a well-established science program is currently underway at the Sanford Underground Research Facility (SURF). The unique underground environment at SURF allows researchers to explore a host of important questions regarding the origin of life and its diversity, mechanisms associated with earthquakes and a number of engineering topics. A deep underground laboratory is also where some of the most fundamental topics in physics can be investigated, including the nature of dark matter, the properties of neutrinos and the synthesis of atomic elements within stars.

SURF is being developed in the former Homestake gold mine, in Lead, South Dakota [1, 2, 3]. Barrick Gold Corporation donated the site to the State of South Dakota in 2006, following over 125 years of mining [4], during which time over 600 km of tunnels and shafts were created in the facility, extending from the surface to over 2450 meters (8000 feet) below ground. The Laboratory property comprises 186 acres on the surface and 7700 acres underground, and the Surface Campus includes approximately 23,500 gross square meters (253,000 square feet) of existing structures. The South Dakota Science and Technology Authority (SDSTA) operates and maintains the Sanford Laboratory at the Homestake site in Lead, South Dakota with management and oversight by Lawrence Berkeley National Laboratory (LBNL).

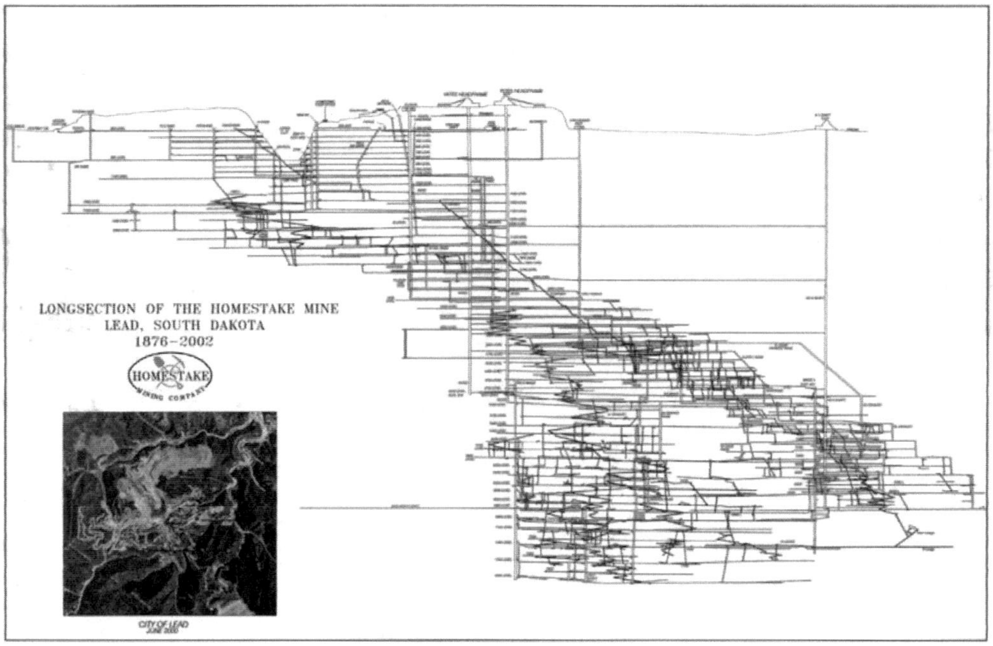

Figure 1. The long section of the former Homestake Gold Mine, in which the dark lines represent vertical shafts and horizontal tunnels projected onto a NW to SE plane. This figure illustrates the 60 underground levels extending to greater than 2450 meters (8000 feet) below ground. For scale, the horizontal length of the projection is 5.2 km. The location of cross section is indicated in the inset.

In 2006, South Dakota philanthropist, T. Denny Sanford, gifted $70M to convert the former mine into a research laboratory and develop a science education facility. With these funds and a strong commitment from the State of South Dakota (appropriations of $42.2M to date), safe access to the underground has been reestablished and experimental facilities have been commissioned and certified for occupancy.

The initial concepts for SURF were developed with the support of the U.S. National Science Foundation (NSF) as the primary site for the NSF's Deep Underground Science and Engineering Laboratory (DUSEL) [5]. With the National Science Board's decision to halt development of a NSF-funded underground laboratory, the U.S. Department of Energy (DOE) now supports the majority of the operation of the facility. Support for experiments at SURF comes from both the NSF and DOE as well as other agencies such as the USGS and NASA. Elements of the Homestake DUSEL Preliminary Design Report [6] continue to be useful as the feasibility for portions of the original plan are investigated.

2. Facility Operations Infrastructure

Maintenance and operation of key elements of facility infrastructure enables safe access underground. Transportation of personnel and materials underground is accomplished using the two primary shafts, the Ross Shaft and the Yates Shaft. Pairs of hoists near both the Ross and Yates shafts move personnel and rock conveyances through the respective shafts. Pumping stations in the Ross Shaft allow ground water to be pumped to surface. Underground ventilation is provided by the Oro Hondo fan as well as the fan at #5 Shaft, which bring fresh air underground via the Ross and Yates Shafts.

A key feature of the Sanford Laboratory is the capacity for redundancy. Redundant power, optical fiber and ventilation air are brought underground via both the Ross and Yates Shafts. Multiple emergency egress options are provided by Ross and Yates shaft (and separate compartments therein) as well as by ramp systems that connect numerous underground levels.

Initial rehabilitation of the surface and underground infrastructure focused on the Ross shaft that provides access to the majority of underground utilities, including the pumping system used to remove water from the mine. Contracted work in the Ross shaft started July 2008 and was completed by October 2008, after which SDSTA personnel continued with maintenance and renovations, including the removal of unused legacy piping (12 km) and power/communication cables (2 km). Started in November 2008, the Yates shaft initial renovation was completed in May 2012 with the installation of a new personnel conveyance and emergency braking system. The Ross Shaft provided primary underground access from the start of re-entry in 2007 until the Yates shaft was ready in May 2012. Personnel and materials are transported underground mainly using the Yates cage, which has dimensions 1.4 m wide × 2.7 m tall × 3.8 m long and has a maximum load capacity of 4540 kg (10,000 lbs); certain loads with widths of up to 1.7 m can be transported beneath the cage. The Yates cage schedule accommodates three shifts per day for science personnel, providing 24-hour access as needed. The maximum underground occupancy at SURF is determined by the number of personnel that can access a safe location in one hour. Based on the current configuration of the various conveyances, the current underground occupancy limit is 72 persons.

In order to provide increased capacity to support the construction and operation of future experiments, state and private funds have been allocated to perform extensive renovations in the Ross Shaft. Refurbishment of the Ross Shaft began in August 2012, and over 670 meters (2200 feet) of new steel and associated ground support has been installed as of the end of December 2014. A total of 1572 m (5159 ft) of new steel sets will be installed as part of the Ross shaft upgrade project, which is expected to be completed by mid-2017. Once the Ross shaft is completed, a renovation of the Yates Shaft will be scheduled.

The average ground water inflow into the underground workings is approximately 730 gpm. After pumping ceased in June 2003 [7, 8], the mine filled with water until a high-water mark of 1381 meters (4530 feet) below surface was reached in August 2008. Sustained pumping resumed in June 2008, dropping the water level below the 4850-foot level (4850L) by May 2009, after being flooded for an estimated 16 months. Since April 2012 the water level is being maintained around the 1830-meter (6000-foot) level below surface. While the potential for accommodating deeper access exists, without a funded mandate to develop laboratory space below the 4850L, there is benefit in terms of cost and safety for maintaining the water level around the current level. If pumping now stopped, it would take an estimated 12–18 months to impact infrastructure on the 4850L.

A deep-well pump was installed July 2010 and is currently located about 1958 meters (6424 feet) below surface in #6 Winze, which extends from the 4850L to the deepest areas of the facility. Permanent stations employing 700-horsepower pumps are located on the 1250, 2450, 3650 and 5000 Levels in the Ross shaft. Water received at the surface Waste Water Treatment Plant (WWTP) is combined with Homestake-Barrick water and treated to remove iron that has leached from the mine workings and trace amounts of ammonia. The discharge capacity from the WWTP is roughly 2000 gpm using biological and sand-filter technologies.

Single-mode fiber optics cable is deployed throughout the facility and current network hardware provides inter-campus communication at 100–1000 Mbps. Redundant connections exist to the outside world, including commodity internet ("Internet 1") at 1 Gbps and research internet ("Internet 2" via the state Research, Education and Economic Development (REED) network) at 1 Gbps, which can be expanded to 100 Gbps with appropriate hardware upgrades. Redundant fiber pathways connect the surface to the 4850L via the Ross and Yates Shafts, and work in underway to ensure that the core network equipment is protected by uninterruptible power supplies and in most cases generator power. Researchers are making good use of the network infrastructure as illustrated in Table 1. In 2014, the average daily throughput was approximately 710 GB.

Table 1. Network traffic value increasing over time, indicating the evolution of science activities at SURF. The majority of the data transfers to date have been associated with the LUX experiment. "Internet 2" was used to route approximately 77% of the data.

Year	Total Data Traffic (TB)
2009	2.6
2010	5.4
2011	12
2012	22
2013	249
2014	259

3. Surface Science Facilities

In addition to the considerable underground extent at SURF, surface facilities exist at both the Yates and Ross surface campuses to facilitate science activities, including administrative support and office space, communication systems for education and public outreach, the WWTP for handling and processing waste materials and a warehouse for shipping/receiving. A number of spaces are currently being used for storage as well as experiment preparation and construction activities. Figure 2 shows the extent of the surface property.

While there are many surface amenities, three main facilities directly serve science needs: the Core Archive, the Sawmill and the Surface Laboratory, all of which are located at the Yates surface campus. Other surface buildings could be modified or renovated to meet specific needs.

3.1. Core Archive

Donated by Homestake-Barrick, SURF is the steward of 39,760 boxes of drill core rock[1], which corresponds to 2688 drill holes with a total length of approximately 91 km. Homestake core holes extend to 3290 meters (10,800 feet) below surface. An additional 1646 m of core were added to the collection in Fall 2009 as part of the geotechnical investigations on the 4850L for DUSEL, and a further 770 m of core have also been collected for geotechnical investigations for the LBNF project in April 2014.

The SD Geological Survey has assisted with the development of an online database that so far includes 58,000+ entries, representing 1740 drill holes.

3.2. Sawmill

A former sawmill is currently being used by SURF personnel as well as researcher groups for various tasks. For instance, the MAJORANA collaboration is performing basic preparation, staging and construction activities in that facility.

[1] Diamond drilling produces cylinders of rock of various diameters called drill core.

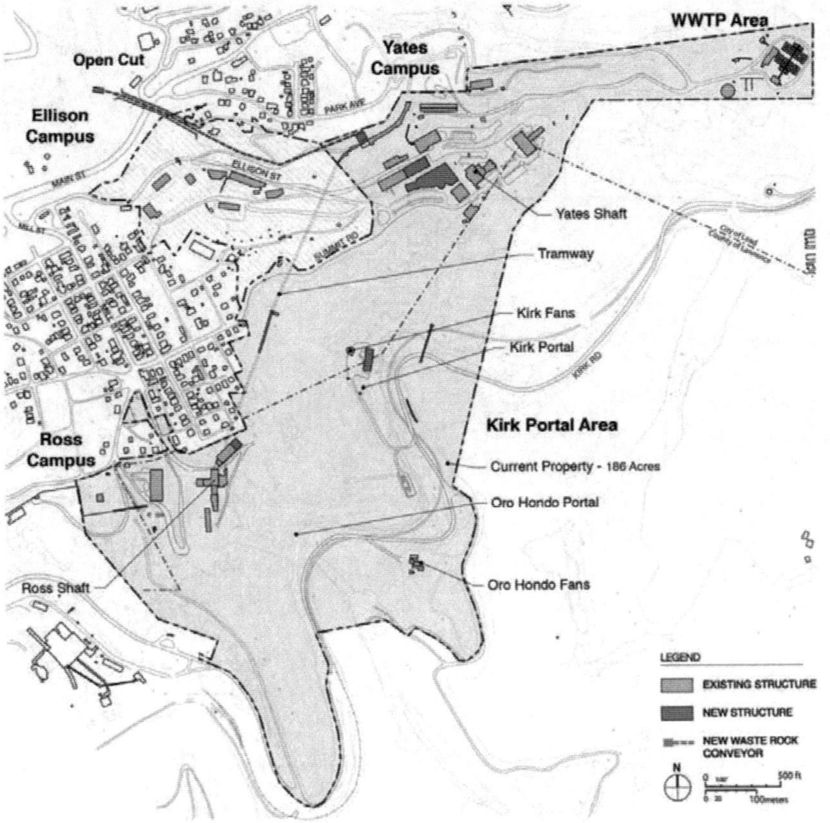

Figure 2. A plan view of the surface campus at SURF, which comprises 186 acres and includes approximately 23,500 m^2 (253,000 ft^2) of existing structures. Locations of the Ross and Yates shafts are indicated as is the main ventilation fan (Oro Hondo) and the Waste Water Treatment Plant (WWTP).

3.3. Surface Laboratory

Renovations were undertaken in 2009 in order to transform a former warehouse into a laboratory. Construction was completed in early 2010, resulting in approximately 190 m^2 of lab space in the top-most level of a four-story building as shown in Figure 3. The facility includes a cleanroom (5.6 m $\times$ 6.6 m with a 2.7 m ceiling height) and corresponding dedicated air handling and filtration system as well as a tank that can be used as a water shield (2.8 m diameter $\times$ 4 m high). The tank is installed in a recessed shaft in the center of the laboratory space. The laboratory was initially designed to meet the needs of the LUX experiment, but is now used to support multiple research groups. Additional renovations to accommodate the LUX-ZEPLIN (LZ) experiment are expected to begin as early as Summer 2015.

4. Underground Science Facilities

A number of levels required to support facility operations infrastructure can also accommodate research activities. However, the main infrastructure for the support of science activities has been developed on the 4850L with formal campuses located near both the Ross and Yates shafts.

Figure 3. Surface Laboratory. The recessed area with the tank is under the grating in the foreground and the cleanroom is located in the background. A hoist can be installed on the blue I-beam structure.

4.1. 4850L Ross Campus

The 4850L Ross Campus includes the Ross Shaft and #6 Winze and encompasses a set of four existing excavations that were used as maintenance shops during mining activities. These former shops afford an economical means to implement experiments or other equipment in a timely manner. The layout is shown in Figure 4, including the location of the current electrical substation[2] and generators. The two western shop areas are currently in use, whereas the two eastern shops are presently undergoing renovations.

Table 2. Footprint areas and volumes for the 4850L Ross Campus spaces. The volume for the NE area is derived from laser-scan data.

Ross Campus Location	**Area (m^2)**	**Volume (m^3)**
NW (MAJORANA Electroforming)	184	504
NE (BHUC)	297	707
SE (CASPAR)	137	376
SW (Refuge Chamber)	121	332
Total	**739**	**1919**

The NW shop area was renovated starting in December 2009 so that a cleanroom (3.7 m wide × 12.2 m long, with an overall height of 3.1 m and an interior ceiling height of 2.5 m) for producing ultra-pure copper could be installed to support the MAJORANA experiment schedule.

[2] The capacity of the current electrical substation near the Ross shaft is sufficient for the planned new developments at the Ross Campus. LBNF will require a separate dedicated substation.

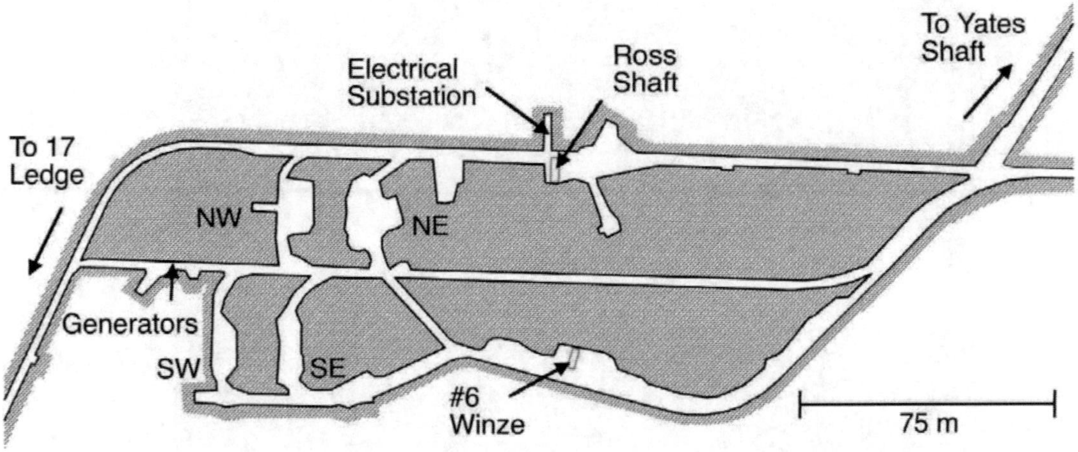

Figure 4. 4850L Ross Campus. Four existing excavations are labeled: NW, NE, SE, SW. The two western shop areas are currently in use, whereas the two eastern shops are presently undergoing renovations.

Figure 5. The MAJORANA electroforming laboratory at the 4850L Ross Campus. The cleanroom is located on the right-hand side.

After several phases of development, the laboratory area, complete with fire suppression system, was available for occupancy in March 2011 as shown in Figure 5. Production copper electroforming operations for the MAJORANA DEMONSTRATOR began in July 2011 and are expected to continue through Spring 2015.

Coincident with the start of the Ross Shaft rehabilitation in the fall of 2012, the SW area was converted into a safety Refuge Chamber that can accommodate 72 people (current maximum occupancy) for up to 96 hours. It includes air locks at the two entrances, compressed air, CO_2

Figure 6. The Refuge Chamber at the 4850L Ross Campus can accommodate 72 people for 96 hours. Amenities shown include first-aid kits, one of two CO_2 scrubbing units and a bank of compressed air cylinders as well as water and rations.

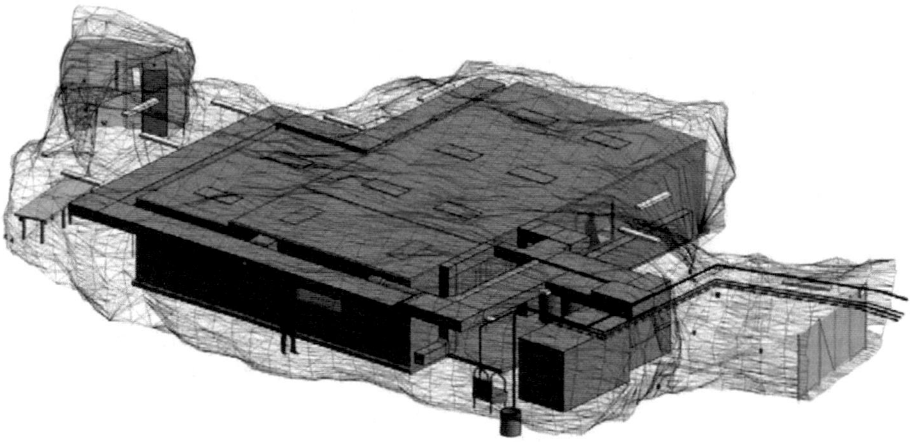

Figure 7. The Black Hills State University Underground Campus at the 4850L Ross Campus. Beneficial occupancy is expected in Summer 2015.

scrubbers, air conditioners, rations and communications. Some of the amenities are shown in Figure 6.

The NE shop will host the Black Hills State University Underground Campus (BHUC) that is intended to support multidisciplinary research from multiple institutions [9]. Installation of additional ground support was completed in December and shotcrete will be applied starting in January, with beneficial occupancy anticipated for Summer 2015. The main feature of the campus will consist of a 268-m^2 cleanroom as illustrated in Figure 7.

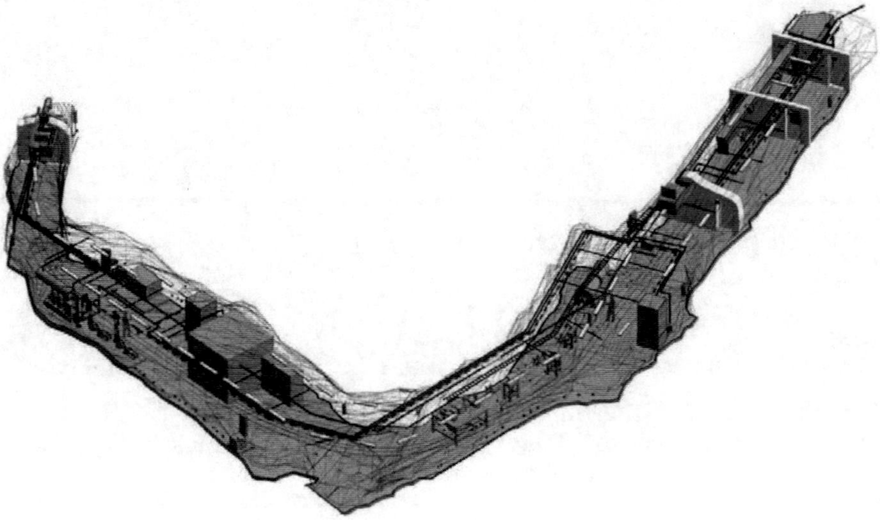

Figure 8. The Compact Accelerator System for Performing Astrophysical Research (CASPAR) will be installed in the SE shop area of the 4850L Ross Campus. An existing accelerator at the University of Notre Dame will be transported to SURF and installed underground starting in Summer 2015.

The Compact Accelerator System for Performing Astrophysical Research (CASPAR) will be installed in a laboratory area comprising 401 m^2, which includes the SE shop area as well as part of an adjoining tunnel. An existing accelerator at the University of Notre Dame will be transported to SURF and installed underground starting in Summer 2015.

4.2. 4850L Davis Campus

A state-of-the-art laboratory complex called the Davis Campus has been constructed at the 4850L near the Yates Shaft. The Davis Campus represents a \$16M South Dakota commitment using state and private funds. New excavation for the Davis Campus took place during September 2009 – January 2011, during which time 16,632 tonnes (18,334 tons) of rock was removed. Rather than being transported to the surface, areas were identified on the 5000L (via the 4850L) for rock storage. Shotcrete was applied in both the Davis Cavern (average thickness 12.7 cm) and the Transition space (average thickness 8.9 cm). Laboratory outfitting began in June 2011 and was substantially complete in May 2012; pictures are shown in Figure 9. A final occupancy certificate was issued in July 2012, which specifies the maximum number of occupants to be 62 persons.

The two main experiments areas are the MAJORANA Lab/Transition space (43 m L × 16 m W × 4 m H) and the Davis Cavern (17 m L × 10 m W × 12 m H), as shown in Figure 10. Quoted dimensions are average values based on post-shotcrete laser-scan data. The Davis Campus footprint consists of 3017 m^2 (32,478 ft^2) total space as indicated in Table 3. While space is at a premium at the Davis Campus, two cutouts exist outside the clean space offering footprints of 30–33 m^2, with an average ceiling height of 3.7 m. These spaces could be made available to groups with modest equipment needs.

4.2.1. Davis Campus Monitoring Monitoring is conducted for a variety of gases at the Davis Campus. Due to the use of cryogens in various sections of the laboratory, a total of nine oxygen

Figure 9. Pictures of the Davis Campus in May 2012 after outfitting had finished: (left) MAJORANA DEMONSTRATOR Detector Room, (right) water shielding tank in Lower Davis for the LUX detector. A considerable amount of equipment has since been installed by both collaborations.

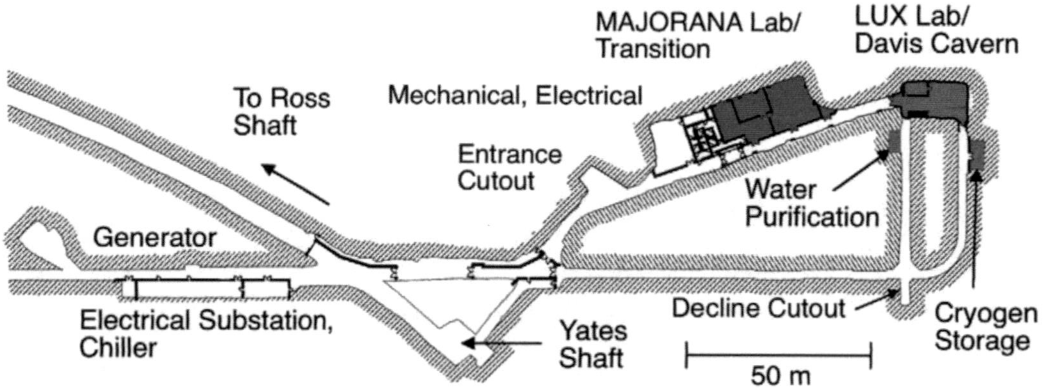

Figure 10. 4850L Davis Campus showing the main laboratory spaces and the proximity to the Yates Shaft, which is approximately 180 m from the main laboratory entrance. MAJORANA spaces are shown in blue whereas the LUX areas are red. The Davis Cavern offers two floors of laboratory space for LUX.

sensors are installed throughout the Davis Campus. Furthermore, a half-dozen portable gas testers sensitive to CO, H_2S, O_2 and CH_4 are also available throughout the campus. A fixed carbon monoxide monitor is located near the Yates Shaft, by the air intake for the Davis Campus. Separate heat and smoke sensors are installed in all areas of the Davis Campus.

A general facility alarm is activated throughout the Davis Campus in the event that any of the following occur: low oxygen, heat or smoke detected, fire suppression water is flowing or a manual pull station is activated. Currently, regular evacuations exercises are held biweekly, with more comprehensive scenarios conducted quarterly.

An array of extensometers is installed to monitor ground motion and convergence at the Davis Campus – data collected so far indicate nothing unexpected. Geotechnical interpretations were performed using drill core and other rock mechanics studies [10].

Table 3. Footprint areas and volumes for the 4850L Davis Campus spaces. The volumes are derived from post-shotcrete (where appropriate) laser-scan data with finished flooring.

Davis Campus Location	Area (m^2)	Volume (m^3)
LUX	375	1976
MAJORANA DEMONSTRATOR	300	1279
Low-Bkgd Counting	33	140
Common Science	244	980
Other/R&D Science	63	253
Facility Infrastructure	508	2063
Access Drifts	1494	4664
Total	**3017**	**11,354**

4.2.2. Davis Campus Services Services provided throughout the Davis Campus include fire sprinklers, potable and non-potable (industrial) water, lighting (including emergency lighting), ventilation and air conditioning. A building management system provides controls throughout the Campus. Cooling is provided with two redundant 50-ton (633-MJ) chillers supplying chilled water to three air handling units that provide ventilation to separate campus spaces. Chilled water and a facility air compressor are also available for experiments to connect equipment directly. A dedicated 1500 kVA substation provides sufficient capacity for the experiment and facility needs, with margin for future expansion (e.g., LZ and any scaling of the MAJORANA DEMONSTRATOR). Emergency power for lighting is provided with batteries in the lighting system to provide immediate light, while a standby diesel generator near the campus provides up to 24 hours of power to all safety systems in the campus. This includes water pumps in the nearby Yates shaft to prevent water from rising into the campus spaces. Three separate air handling units provide multi-stage air filtration for the Davis Campus clean spaces and establish relative pressure balances to help maintain appropriate relative levels of cleanliness. Starting in July 2014, the intake air for the Davis Campus clean area air handling system was pre-processed using a dedicated dehumification system.

4.2.3. Davis Campus Cleanliness Cleanliness at the Davis Campus is maintained through a combination of transition zones and protocols as well as dedicated cleaning staff. When supplies and equipment are brought from the dirty space into the clean space, items either need to be suitably enclosed or cleaned in the facility entrance Cart Wash. Personnel typically enter though a series of rooms where outer coveralls and dirty-side gear are removed in favor of corresponding clean items. Laundry services are also available in the clean space. The facility design includes two sets of personnel showers, but so far they have not been necessary.

When the laboratory is occupied, particle counts (>0.5 μm per cubic foot) in all main common areas are below 10,000/ft^3. The MAJORANA collaboration follows additional gowning protocols to achieve ultra-low cleanliness levels in their laboratory spaces. Particle counts have been monitored at the Davis Campus since September 2012, and a summary of cleanliness states throughout various areas is presented in Table 4.

Dust from the air-handling unit filters as well as floor sweepings have been collected and routinely analyzed for more than 6 months. The main activity contributions appear to be from ^{210}Pb and ^{7}Be. For example, *in situ* dust has the following average levels: ~1 ppm U, ~2 ppm Th and 1–2% K [11].

Journal of Physics: Conference Series **606** (2015) 012015 doi:10.1088/1742-6596/606/1/012015

Table 4. Median particle count levels in various Davis Campus laboratory locations during occupied periods. Additional cleanliness protocols are instituted in the MJD areas.

Davis Campus Location	**Particle Count ($0.5\ \mu m/ft^3$)**
Entry/Lockers (MJD)	790
Common Transition Hallway	2320
Common Corridor	3240
Davis Cavern, Lower (LUX)	3400
Davis Cavern, Lower (CUBED)	720
Davis Cavern, Lower (BLBF)	490
Davis Cavern, Upper (LUX)	400
Detector Room (MJD)	100–400

5. Facility Characteristics

A number of specific properties that define and influence capabilities for various underground laboratory spaces are reviewed in the sections below, including the geographic coordinates, geology and rock overburden characteristics as well as radon and radioactivity considerations.

5.1. Coordinates

Initial survey work in 2010 projected the Universal Transverse Mercator (UTM) coordination system (Zone 13 North, NAD83 datum with 2002 Epoch) underground to benchmarks on the 4850L. Projected Easting (x) and Northing (y) coordinates for various underground locations are presented in Table 5. Future surveys in support of the long baseline neutrino initiative will improve on these preliminary efforts.

5.2. Geology and Rock Overburden

When considering the muon flux at key laboratory spaces on the 4850L, six geological formations are important to consider: Grizzly, Flagrock, Northwestern, Ellison, Homestake and Poorman. The Yates Member (Unit) is the lowest stratigraphic unit of the Poorman Formation and is important for developments on the 4850L; other geological formations also have amphibolite manifestations that are accounted for in the analysis below. In addition, tertiary Rhyolite dikes occur throughout the rock mass.

The surface topology varies significantly throughout the laboratory property as shown in Figure 11. Using average rock densities and a 3-dimensional geological model [13] in addition to recent topological surveys, estimates of the rock overburden for various locations are summarized in Table 6. Errors in the vertical rock depth are expected to be modest (± 10 m). To estimate the overburden uncertainties associated with the geological model, rock formation contact boundaries were adjusted for several locations based on the relative information about their locations in order to maximize and minimize the total density-scaled depths. Density errors (1–6%) were also factored in using $\pm 1\ \sigma$ (standard deviations), and contributed the majority of the uncertainty. However, in addition to simply quantifying the measurement error, the uncertainties also characterize density variations due to differences in mineralogy in a given rock formation. As a result of the combination, the average uncertainty in the water-equivalent overburden values was estimated to be $\sim$4% for most of the locations; the overburden error for the 800L location is lower at roughly $\sim$2.5%. In addition to the vertical overburden depth values, the rock density averaged over a 45-degree cone (determined by sampling the model at several

Table 5. Coordinates estimates for various locations of interest. Both UTM and geographical coordinates are presented. U.S. survey feet are converted to meters.

Location	**UTM Coordinates**		**Geographic Coordinates**	
	Easting (m)	**Northing (m)**	**Latitude**	**Longitude**
4850L Davis Campus				
LUX Detector	599527	4911873	44° 21' 11.88" N	103° 45' 04.27" W
MJD Detector	599516	4911838	44° 21' 10.75" N	103° 45' 04.77" W
Entrance Cutout	599531	4911781	44° 21' 08.92" N	103° 45' 04.15" W
4850L Ross Campus				
MJD Electroforming	598939	4911087	44° 20' 46.70" N	103° 45' 31.33" W
BHUC	598962	4911085	44° 20' 46.65" N	103° 45' 30.31" W
CASPAR	598939	4911045	44° 20' 45.36" N	103° 45' 31.39" W
4850L Other				
LBNF (10 kt Cavern)	599236	4911041	44° 20' 45.07" N	103° 45' 17.96" W
LBNF (30 kt Cavern)	599223	4910960	44° 20' 42.44" N	103° 45' 18.62" W
Expt Hall (Generic)	599437	4911408	44° 20' 56.88" N	103° 45' 08.66" W
Other				
800L (Muon site [12])	598928	4911221	44° 20' 51.07" N	103° 45' 31.74" W
2000L (Muon site [12])	598764	4912928	44° 21' 46.47" N	103° 45' 37.98" W

discrete points through the cone volume) is presented for various underground locations as a way of characterizing the effective overburden over a broader extent. Geochemical composition data for various formations are available from sources such as the USGS [14] and others.

5.3. Radioactivity and Radon

SURF strives to provide the lowest possible radioactivity environment for experiments hosted within the facility. This commitment had been integrated into the site preparation process from the early days of the facility design, and carried over to the realization of the 4850L Davis Campus laboratories. These efforts include site and environmental characterization including rock radioactive measurements, use of low-radioactivity construction materials, and regular monitoring of environmental factors, including airborne radon.

The natural abundance of U/Th/K in Homestake rock is generally low, especially in certain geological formations such as the Yates Amphibolite, which have been measured to contain sub-ppm levels of U/Th. Samples from other formations such as the Rhyolite intrusions can be 30–40× higher as illustrated in Table 7. While not as low as the host Yates Amphibolite rock, the lowest-activity construction materials for the Davis Campus were selected from a large number of samples [15, 16, 17].

Long-term underground radon data have been collected since 2009 and at the Davis Campus since July 2012, shortly after outfitting was completed. Aside from the normal facility ventilation that brings fresh air underground via the Ross and Yates shafts, no extraordinary steps have been taken to mitigate underground radon levels. Some research groups employ a type of purging system to reduce radon locally for their equipment. As seen at other underground laboratories,

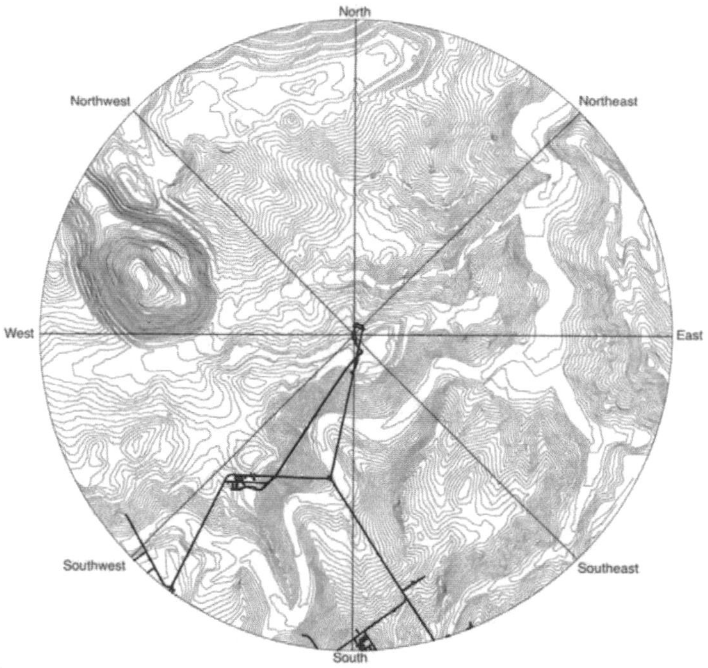

Figure 11. Topological contours for the opening of a 45-degree cone overlaid on top of an outline of the 4850L (shown in blue), centered on the Davis Campus. Various cardinal coordinates are indicated with red lines. The Open Cut is a prominent feature on the surface toward the west. Contour lines indicate 6-meter (20-foot) elevation changes.

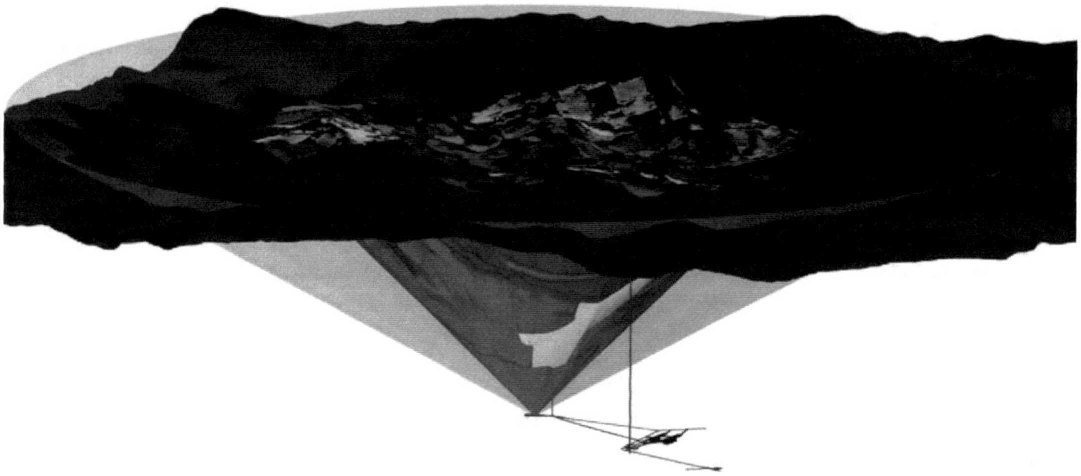

Figure 12. 3D geological model showing rock cone above the 4850L centered on the Davis Campus as well as the surface topology. The colored section represents different rock formations in the 45-degree cone; the gray outline represents the extent of a 60-degree cone. The vertical lines show the locations of the Ross and Yates shafts.

Table 6. Rock overburden estimates for various underground locations at SURF, using a 3-dimensional geological model [13] and including depths scaled by density to give the meters of water equivalent (m.w.e.). The average rock density is based on a 45-degree cone centered above the specific site. The data reflect considerable variations in surface topology as well as geology. Properties related to the locations of recent muon flux measurements are also included. Uncertainties on the rock depth values are estimated to be ±10 meters, whereas the estimated uncertainties in the density-scaled overburden values are ~4%; the overburden error for the 800L location is lower at roughly ~2.5%.

Location	**Vertical Rock Overburden (m)**	**(m.w.e.)**	**Average Cone Density (g/cm^3)**
4850L Davis Campus			
LUX Detector	1466	4210	2.85
MJD Detector	1477	4260	2.86
Entrance Cutout	1494	4300	2.85
4850L Ross Campus			
MJD Electroforming	1503	4290	2.81
BHUC	1503	4380	2.83
CASPAR	1499	4170	2.81
4850L Other			
LBNF (10 kt Cavern)	1384	3910	2.82
LBNF (30 kt Cavern)	1375	3840	2.82
Expt Hall (Generic)	1189	3390	2.80
Other			
800L (Muon site [12])	283	770	2.73
2000L (Muon site [12])	624	1700	2.74

a seasonal dependence is becoming apparent in the Davis Campus trends, with higher radon levels observed during the summer months May through September. The total average radon concentration over the monitoring period (883 days) at the Davis Campus is approximately 310 Bq/m^3 with a low baseline of 150 Bq/m^3 during the winter months. Brief excursions above 1000 Bq/m^3 have been observed, typically correlated with known events such as ventilation fan maintenance or when the Yates and/or Ross shafts are partially covered.

Other efforts to characterize physics backgrounds (eg., muons, neutrons, gamma rays) in various underground areas were carried out by various research groups. Historically, the differential muon flux was measured by collaborators on the Homestake chlorine experiment [18]. More recently, since the mine was reopened as a laboratory, other groups have gathered data over several years [12, 19, 20]. Results from recent neutron flux measurements collected on the 4850L are expected in an upcoming publication as are results related to the cosmic ray muon flux measured at the Davis Campus.

Table 7. Partial U/Th/K assay results for relevant SURF rock samples as well as key construction materials used at various laboratories [15, 16, 17].

	Uranium (ppm)	Thorium (ppm)	Potassium (%)
4850L Davis Campus			
Yates Amphibolite Fm (Majority)	0.22	0.33	0.96
Rhyolite Dike	8.75	10.86	4.17
Shotcrete – Low Activity	1.52	2.17	0.55
Shotcrete – Standard	2.00	3.35	1.23
Shotcrete – Finish Coat	1.62	3.08	0.79
Masonry Blocks	2.16	3.20	1.23
4850L Ross Campus			
Poorman Fm (Majority)	2.58	10.48	2.12
Siderite/Grunerite Ironstone	0.41	2.30	0.25
Shotcrete – Standard	Expect similar to Davis Campus		
Shotcrete Surface Coating	Under investigation		

6. Current Science Program

Science efforts that started in 2007 during re-entry into the facility have grown steadily over the years. Building on the legacy of the Ray Davis chlorine solar-neutrino experiment [21] that began in 1965 at the Homestake Mine, twenty-six research groups have conducted underground research programs at SURF. Measurements have been made and samples collected from elevations ranging from surface to the 5000L, investigating topics in physics, geology, biology and engineering. A total of eighteen research programs are considered active as summarized in Table 8, which includes a brief description of each group as well their respective locations. The list includes the two main physics efforts underway at Davis Campus, as well as the two low-background counters. There has also been significant interest from additional groups across all disciplines.

6.1. Large Underground Xenon (LUX)

The LUX experiment [22] is conducting a direct search for weakly interacting massive particles (WIMPs) using 370 kg (118 kg fiducial) of purified xenon inside an ultra-pure titanium cryostat that is immersed in a water tank containing 272,550 liters (72,000 gallons) of purified water. Following collisions in the liquid xenon volume, a strong electric field moves ionization electrons into the gas space at the top of the detector, and a total of 122 PMTs collect scintillation light from interactions in both xenon volumes.

Members of the LUX collaboration have been active onsite at SURF since December 2009, when preparations began for detector assembly at the Surface Laboratory. After completing detector assembly and operational testing on surface, LUX began their transition underground to the Davis Campus in May 2012. With strong support from SURF personnel, the fully-assembled LUX detector was successfully transported underground from the Surface Laboratory in the Yates cage and into the Davis Cavern – stringent limits on acceleration and tilt were not exceeded. The LUX space at the Davis Campus includes a cleanroom and a control room, which are separate from the main lab space. The SURF water purification system and a storage area for liquid nitrogen are located near the LUX equipment but outside the clean environment.

Table 8. Current scientific research programs at the Sanford Laboratory. Locations in bold indicate current installations or the subject of current focus; those in italic are imminent.

Experiment	Description	Locations	References
Physics			
Large Underground Xenon (LUX)	Dark matter using Xe	Surface, **4850L**	[22, 23]
MAJORANA DEMONSTRATOR	Neutrinoless double-beta decay using Ge	**Surface**, 800L, **4850L**	[24, 25]
Center for Ultra-low Bkgd Experiments in the Dakotas (CUBED); also Bkgd Characterization	Low-bkgd counter, isotope separation; also bkgds: muon, neutron, gamma, radon	**Surface**, 800L, 2000L, 4550L, **4850L**	[12, 19, 20, 26]
Berkeley Low-Bkgd Facility	Low-bkgd counter	**4850L**	[27]
Compact Accelerator System for Performing Astrophysical Research (CASPAR)	Neutron bkgds	4100L, 4850L	[28]
Long-Baseline Neutrino Facility (LBNF)	Cleanliness tests	Surface, 4850L	[29]
Deep Underground Gravity Laboratory (DUGL)	Seismic characterization for gravity-wave research	**Surface, 4850L,** 300L, 800L, 1700L, 2000L, 4100L	[30, 31]
Geology			
Geoscience Optical Extensometers and Tiltmeters (GEOX^{TM})	Optical fiber applications, tiltmeters	**2000L, 4100L, 4850L**	[32, 33, 34, 35], [36]
USGS Hydrogravity	Local gravity for water tables, densities	**Surface, 300L, 4100L, 4850L**	[37]
Petrology, Ore Deposits, Structure (PODS)	Core archive and logs, geologic mapping	Surface, 800L	[38, 39, 40]
Transparent Earth	Seismic monitoring/ properties, melt studies	2000L, **4100L 4850L**	[41, 42, 43]
Biology	(Samples collected informally have also resulted in publications [44])		
Biodiversity (BHSU)	Microbiology	Surface, 300L, 2000L, **4100L**, 4550L, **4850L**	[45, 46, 47]
Biofuels (SDSMT)	Biofuels	4550L, **4850L,** 5000L	[48, 49, 50, 51], [52, 53, 54, 55]
Lignocellulose (SDSU)	Biofuels	1700L, **4850L**	[56, 57]
Syngas (SDSMT)	Biofuels	**4850L**	[58, 59, 60]
Life Underground – NASA Astrobiology Institute	Water in drill holes, geomicrobiology	Surface, **800L, 4850L**	[61]
Engineering			
Xilinx, Inc.	Chip error testing	*4850L*	

The LUX collaboration published results in October 2013 from their first underground data run [23] and started a longer run (nominally 300 live days) in October 2014. LUX is expecting to complete their experimental program in 2016.

6.2. MAJORANA DEMONSTRATOR (MJD)

The MAJORANA collaboration is investigating neutrinoless double-beta decay using a detector called the DEMONSTRATOR consisting 40 kg of germanium in two ultra-pure copper cryostats and a compact shield constructed from lead, copper and high-density polyethylene. MAJORANA expects to use up to 30 kg of enriched ^{76}Ge. The main technical goal of the DEMONSTRATOR is to confirm that the ambitious purity requirements for a tonne-scale detector are achievable [24, 25]. The group also plans to test a controversial claim for the detection of neutrinoless double-beta decay.

In order to satisfy the very low background requirements, the MAJORANA collaboration has set up a laboratory at the 4850L Ross Campus to produce the world's purest electroformed copper, which is then machined at a dedicated machine shop at the Davis Campus. The MJD electroforming process begins with minimally-processed oxygen-free high conductivity copper nuggets (99.999%), with typical properties: U = 1.25 μBq/kg, Th = 1.10 μBq/kg and yield strength of $\sim$50 MPa. MJD electroformed copper averages U = 0.17 μBq/kg, Th = 0.06 μBq/kg with yield strength 70–100 MPa (work hardened). The average growth rate for electroformed copper is relatively slow at approximately 1 mm/month, and MJD expects to produce roughly 2860 kg of the material for the DEMONSTRATOR.

The MAJORANA collaboration began work onsite starting in November 2010 with deliveries of equipment (including the first germanium detectors) and then with preparations for the electroforming cleanroom. Production of electroformed copper began in July 2011 and about eight months later in March 2012 the group started to move equipment into the Davis Campus. The majority of the detector shield is complete, including 3400 cleaned and stacked Pb bricks and some of the inner copper sheets. MJD is currently collecting background data from germanium crystals in a prototype (commercial-copper) cryostat. Completed assembly and the start of commissioning for the first production module with 20 kg of high-purity germanium (mostly enriched in ^{76}Ge) is expected in early 2015, with the second production module expected to be completed in late 2015. The full MAJORANA experimental program with both production cryostats could extend into 2019.

6.3. Center for Ultra-low Background Experiments in the Dakotas (CUBED)

Initially created as a South Dakota Governor's Research Center administered through the University of South Dakota, CUBED involves physics researchers and others from the majority of universities in the state [26]. In addition to increasing the South Dakota academic involvement in SURF research, the center maintained a focus on areas of interest congruent with planned SURF experiments, such as crystal growth, low-background counting and stable isotope separation.

CUBED has set up a low-background counting laboratory in a dedicated room in the Lower Davis Cavern of the 4850L Davis Campus. The system allows direct-gamma counting using a 1.3-kg high-purity n-type germanium crystal (60% relative efficiency), resulting in 0.06–0.1 ppb sensitivities to U/Th as summarized in Table 9. Installation of the germanium counter began in April 2013 and was finished a year later in April 2014 when the complete shield was established: 72 oxygen-free high conductivity copper bricks, a stainless steel enclosure for purging radon using boil-off nitrogen and approximately 500 lead bricks that were cleaned and coated in borated paraffin. Assays of community samples began in October 2014. A subsequent reconfiguration of the detector shield carried out in December 2014 has resulted in a further 25% reduction in the rate of background events compared to the original performance. The CUBED low-background counter is expected to move to a separate dedicated cleanroom at the 4850L Ross Campus in Summer 2015.

Another CUBED project, the isotopic separation and ultra-purification (ISUP) project, is underway at the Surface Laboratory, where the separation of stable isotopes is being investigated. Initial tests are being performed with CO_2, with plans to also study N_2, Ar, Kr and Xe. Installation began in May 2014 with commissioning and full operation in September 2014. Studies are expected to be completed in May 2015.

A main CUBED focus involves crystal growth, and SURF personnel have provided a cost estimate to develop a laboratory to support underground crystal growth. Funding is being sought.

6.4. Berkeley Low-Background Facility (BLBF)

After operating since the 1980s, LBNL counting operations at the Oroville Dam[3] ended in February 2014. The site was decommissioned and the HPGe detector and shield materials were shipped to SURF in March. Starting in May, the detector and shielding were installed at a dedicated laboratory within the 4850L Davis Campus, next door to the CUBED low-background counter. After a brief period of commissioning and background testing in June, the counter was back online for receiving samples in July [27]. In total, the implementation at SURF took less than 6 months. Since July, eleven community samples have been counted on behalf of the MAJORANA and LZ collaborations.

The BLBF counter is a 2.1-kg p-type coaxial germanium crystal (85% relative efficiency), resulting in 0.01–0.3 ppb sensitivities to U/Th (see Table 9). Compared to the original Oroville installation, background rates measured at SURF indicate a performance improvement of approximately 30%, mainly due to shield configuration improvements. The BLBF low-background counter is expected to move from the Davis Campus to a separate dedicated cleanroom at the 4850L Ross Campus in Summer 2015.

Table 9. Low-background counter sensitivities for a sample of order ~1 kg and counting for approximately two weeks.

	Uranium		**Thorium**		**Potassium**	
	(ppt)	**(mBq/kg)**	**(ppt)**	**(mBq/kg)**	**(ppb)**	**(mBq/kg)**
CUBED	60	0.7	100	0.4	25	0.7
BLBF	10	0.1	30	0.1	20	0.6

6.5. Experiment Implementation Process

SURF has developed a formal process for implementing experiments [62]. Requirements include an experiment planning statement, a memorandum of understanding, appropriate insurance as well as a decommissioning plan. If necessary, financial responsibilities for both SURF and the experiment are documented in a general services agreement. Safety reviews are commensurate with the scale of the project. SURF personnel are available to serve in various roles, including acting as a guide in the underground environment, providing logistical support as well as participating in the work planning process.

[3] The Oroville Dam was the original site of the UCSB/LBL double-beta decay experiment.

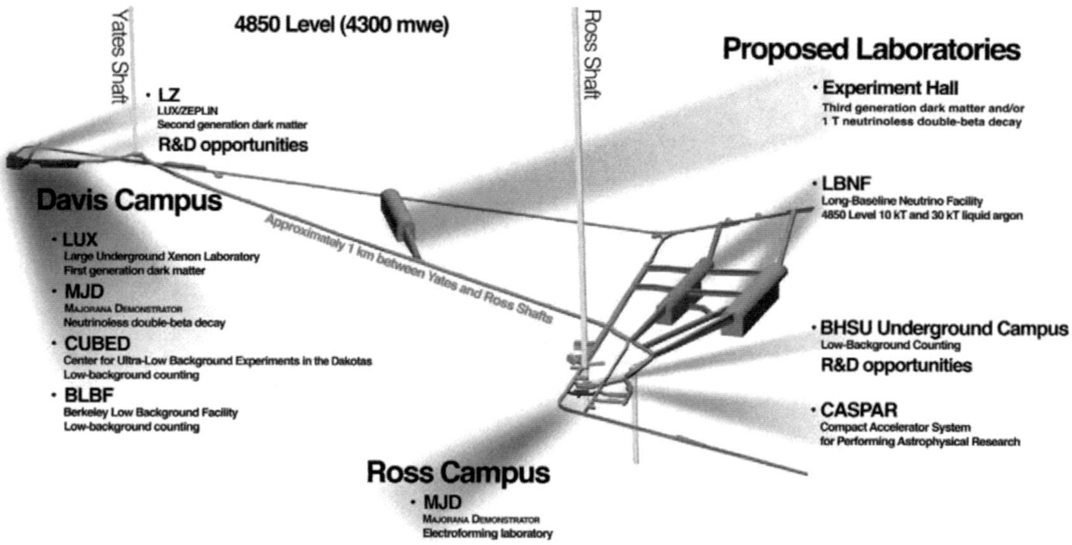

Figure 13. The 4850L of SURF highlighting the existing and proposed experiments.

7. Future Science

The future offers many scientific opportunities for underground science that can be pursued at SURF. Both the nuclear physics and high-energy physics communities are developing strategic plans for the future of their respective fields, and projects that are planned for SURF figure prominently [63]. Near-term expansions are underway and there exist options to significantly expand the facility footprint to accommodate additional endeavors. The locations of existing and future experiments are shown in Figure 13.

7.1. Black Hills State University Underground Campus (BHUC)

As mentioned previously, Black Hills State University is developing an underground campus at the 4850L Ross Campus, including a cleanroom that will host a variety of research activities. Approximately 75% of the BHUC cleanroom space is envisioned to be a low-background assay laboratory operated at Class 1000, while the remaining 25% of the laboratory will be reserved for research that does not require a high level of cleanliness. Space in the cleanest laboratory area dedicated to performing low-background assays is expected to house 8–9 instruments, possibly including more than a half-dozen high-purity germanium detectors and the associated shielding and access systems as shown in Figure 14. The area outside the cleanroom will be used for storage as well as other research activities.

7.2. Compact Accelerator System for Performing Astrophysical Research (CASPAR)

An underground facility is under development at the 4850L Ross Campus to study the stellar nuclear fusion reactions responsible for the production of elements heavier than iron, especially via the slow neutron-capture nucleosythesis process (s-process). The Compact Accelerator System for Performing Astrophysical Research (CASPAR) collaboration seeks to study (α,n) reactions at stellar energies relevant for helium and carbon burning [64]. CASPAR will use an existing 1-MV Van de Graff accelerator from the University of Notre Dame capable of producing high-intensity ($\sim$150 μA) proton and helium beams with energies between 200 keV and 1 MeV. A specialized windowless gas target system for ^{22}Ne and ^{13}C is also being developed at the

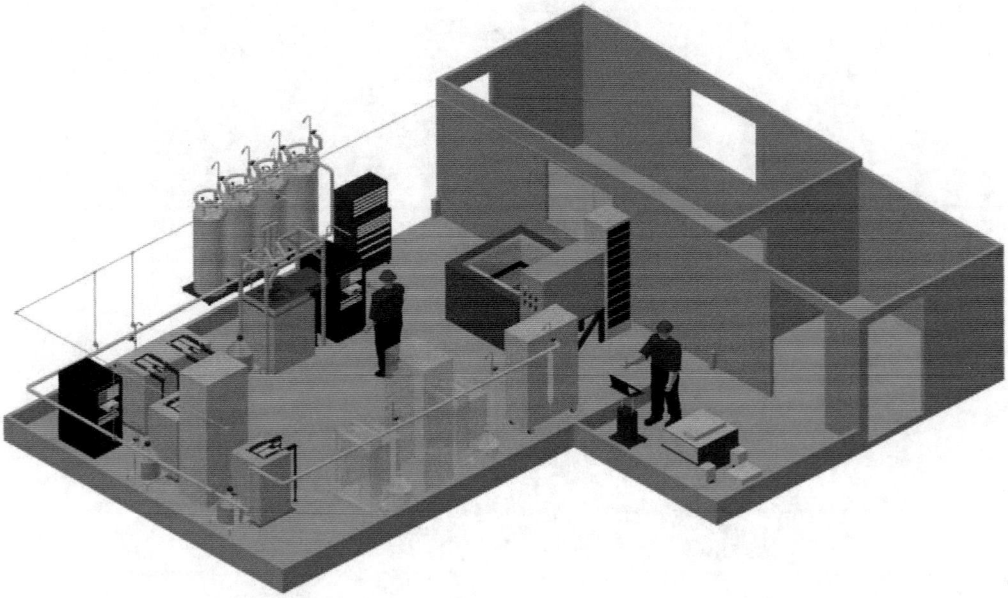

Figure 14. Schematic of the Black Hills State University Underground Campus at the 4850L Ross Campus, showing a nominal set of low-background assay instruments and a liquid nitrogen supply and purge system. The rooms on the right-hand side are intended to be used to conduct biology and geology research.

Colorado School of Mines. The South Dakota School of Mines and Technology will operate the facility.

The initial CASPAR scientific program is expected to take 5–8 years to complete, with a broader program that is expected to last for over a decade. Potential future federal funding could allow for expansion to include additional elements that were originally planned for the Dual Ion Accelerators for Nuclear Astrophysics (DIANA) project [65].

7.3. LUX-ZEPLIN (LZ)

In July 2014, the LUX-ZEPLIN (LZ) proposal was one of two direct-search generation-2 dark matter experiments selected for funding by the U.S. DOE division of High Energy Physics. The LZ detector will employ approximately 10 tonnes of liquid xenon ($\sim$6 tonnes fiducial or $\sim$45$\times$ LUX fiducial), and will rely on modest upgrades to existing SURF infrastructure, including the Surface Laboratory for detector assembly and underground space at the 4850L Davis Campus where the LUX experiment is currently operating. In particular, no new excavation will be required, and LZ will be able to reuse the water shielding tank used for LUX with the addition of an improved background veto system consisting of gadolinium-loaded linear alkyl-benzene liquid scintillator, as shown in Figure 15. LZ sensitivity is projected to be approximately 100$\times$ better than the anticipated LUX 300-day result (underway), with sensitivity of 2×10^{-48} cm^2 for its full 1000-day exposure.

Detector assembly at the Surface Laboratory could begin as early as mid-2015. Underground production data-taking is scheduled to begin 2019 and will nominally last 5 years.

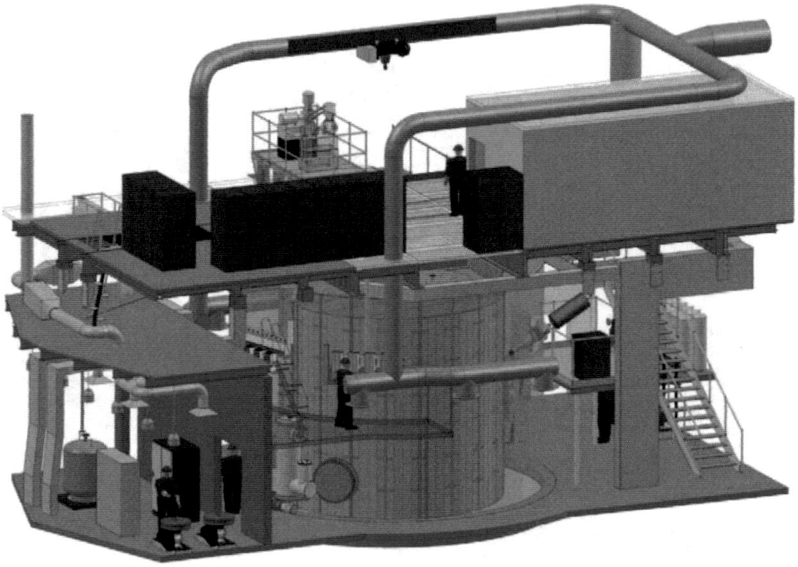

Figure 15. Schematic of the LUX-ZEPLIN experiment at the 4850L Davis Campus. Existing infrastructure is shown such as the ductwork, the water shielding tank (lower center) and the control room (upper right). LZ will require all available space in the Davis Cavern, including rooms (lower left) currently used by the two low-background counters.

7.4. Neutrinoless Double-Beta Decay

The MAJORANA collaboration is investigating performing the next-generation neutrinoless double-beta decay experiment on the 4850L at SURF. If the combination of rock overburden and veto is determined to be sufficient, scaling up to order ∼200 kg from the current DEMONSTRATOR may be possible at the Davis Campus but ultimately a new laboratory may be needed.

7.5. Long Baseline Neutrino Facility (LBNF)

Neutrinos propagated over long distances (> 1000 km) allow for investigations into fundamental physics, touching on such compelling unanswered puzzles as neutrino mass hierarchy and CP violation (matter/antimatter asymmetry) as well as many other topics.

For many years, the U.S. has been developing an experiment called the Long Baseline Neutrino Experiment (LBNE) [66] that is transforming into the Long Baseline Neutrino Facility (LBNF). While the project design has undergone many iterations (consideration of water Čerenkov technology and an initial detector installation on the surface), the latest design consists of a large liquid argon detector (or detectors) located underground at the 4850L of SURF that would observe neutrinos generated approximately 1300 km away at Fermilab using an upgraded accelerator beam (up to 2.4 MW). The detector mass could be staged in various ways (e.g., 10 kt, 30 kt, etc.) to reach a nominal fiducial mass of 40 ktonne (total mass ∼60 ktonne). Some simulations to quantify backgrounds due to cosmic ray muons have already been performed [67].

International partners are required to ensure the success of such an ambitious project, and a process is currently underway to consolidate interested researchers into a single Fermilab-led experiment. The new proto-collaboration is referred to as the Experiment at the Long Baseline Neutrino Facility (ELBNF) and is expected to include the majority of researchers from existing collaborations such as LBNE (90 institutions and 500+ collaborators) and the Long Baseline

Neutrino Oscillation (LBNO) European experiment [68]. A Letter of Intent defining the new collaboration has been drafted [69] and will be presented to the Fermilab Program Advisory Committee (PAC) in January 2015.

Geotechnical studies have already commenced, and the first construction efforts associated with LBNF could begin as early as FY16 when shaft services and some aspects of a waste-rock handling system could be installed. Underground excavation could begin as soon as the Ross shaft rehabilitation is completed, expected to be mid-2017.

7.6. Future Expansion

Space is currently available for modest R&D efforts, including cutouts at the Davis Campus (30–33 m^2) and laboratory space at the BHSU Underground Campus. While it may be possible to utilize a few other existing excavations on the 4850L and other relatively deep levels, most practical existing excavations will soon be occupied. Some spaces that are currently occupied may become available again in 5–10 years.

New excavations of order 11,500 cubic meters (15,000 cubic yards) of rock could be performed prior to the completion of the Ross Shaft rehabilitation; however, once a waste-rock handling system is installed to support the LBNF project very large excavation projections become possible. Based on the DUSEL preliminary design, a generic Experiment Hall is included on Figure 13 as a placeholder to accommodate a tonne-scale neutrinoless double-beta decay experiment and/or a generation-3 dark matter experiment. Nominal dimensions are 20 m W $\times$ 24 m H $\times$ 50–100 m L displacing 45,900–85,600 cubic meters (60,000–112,000 cubic yards) of rock, including access tunnels, over-break and rock swell. Other configurations and locations are also possible.

8. Summary

SURF is a deep underground research facility, dedicated to scientific uses. Research activities are supported at a number of facilities, both on the surface and underground. Two campuses on the 4850L accommodate a number of leading efforts, and in particular the 4850L Davis Campus has been successfully operating for over 2.5 years. The LUX and MAJORANA experiments are well established and there are current capabilities for low-background counting. Many expansion possibilities are on the horizon and a number of key experiments in the U.S. research program are developing plans for installation at SURF.

References

[1] Heise J 2014 *AIP Conf. Proc.* **1604** 331–344 (*Preprint* `arXiv:1401.0861 [physics.ins-det]`)
[2] Lesko K T 2012 *Eur. Phys. J. Plus* **127** 107–118
[3] Heise J 2010 *Nuc. Phys. A* **834** 805c–808c
[4] Mitchell S T 2009 *Nuggets to Neutrinos: The Homestake Story* (Xlibris)
[5] Lesko K T 2013 Why the US needs a deep domestic research facility: Owning rather than renting the education benefits, technology advances, and scientific leadership of underground physics (*Preprint* `arXiv:1304.0402 [physics.ins-det]`)
[6] Lesko K T *et al.* 2012 Deep Underground Science and Engineering Laboratory — preliminary design report (*Preprint* `arXiv:1108.0959 [hep-ex]`)
[7] Murdoch L C, Germanovich L N, Wang H, Onstott T C, Elsworth D, Stetler L and Boutt D 2012 *Hydrogeol. J.* **20** 27–43
[8] Zhan G and Duex T 2010 *Mining Eng.* **64(4)** 64–68
[9] Mount B 2015 Low background counting facility in BHSU Underground Campus at Sanford Lab *Proceedings for this workshop*
[10] Pariseau W G, Tesarik D R and Trancynger T C 2012 *Transactions of the Society for Mining, Metallurgy, and Exploration* **332** 370–388
[11] Thomas K J, Lesko K T and Smith A R 2014 Berkeley low background facility update *Private Communication*
[12] Gray F E, Ruybal C, Totushek J, Mei D -M, Thomas K and Zhang C 2010 *Nucl. Instrum. Meth. A* **638** 63–66 (*Preprint* `arXiv:1007.1921 [nucl-ex]`)

[13] Hart K, Trancynger T C, Roggenthen W and Heise J 2014 *Proceedings of the South Dakota Academy of Science* **93** 33–41 `http://www.sdaos.org/wp-content/uploads/pdfs/2014/roggenthen%2033-41.pdf`
[14] Caddey S W, Bachman R L, Campbell T J, Reid R R and Otto R P 1991 *U.S. Geol. Surv. Bull.* **1857-J** J1–J67
[15] Roggenthen W and Smith A R 2008 White paper: U, Th, K contents of materials associated with the Homestake DUSEL site, Lead, South Dakota *Private Communication*
[16] Smith A R 2007 and updates through 2014 Homestake samples: Results of radiometric analyses at LBNL *Private Communication*
[17] Chan Y D 2012 The low-background construction of laboratories at the 4850-ft level Davis Campus `http://www.sanfordlab.org/lbnl/186`
[18] Cherry M L, Deakyne M, Lande K, Lee C K, Steinberg R I, Cleveland B and Fenyves E J 1983 *Phys. Rev. D* **27** 1444–1447
[19] Thomas K J 2011 *Radon monitoring and the role of iron oxide on radon emanation at the Sanford Laboratory at Homestake in Lead, South Dakota* Master's thesis (unpublished) University of South Dakota Department of Physics
[20] Mei D -M, Zhang C, Thomas K and Gray F 2010 *Astropart. Phys.* **34** 33–39 (*Preprint* `arXiv:0912.0211 [nucl-ex]`)
[21] Cleveland B T, Daily T, R Davis J, Distel J R, Lande K, Lee C K, Wildenhain P S and Ullman J 1998 *Astrophys. J.* **496** 505–526
[22] Akerib D S *et al.* 2013 *Nucl. Instrum. Meth. A* **704** 111–126 (*Preprint* `arXiv:1211.3788 [physics.ins-det]`)
[23] Akerib D S *et al.* 2014 *Phys. Rev. Lett.* **112** 091303 (*Preprint* `arXiv:1310.8214 [astro-ph.CO]`)
[24] Abgrall N *et al.* 2014 *Advances in High Energy Physics* **2014** Article ID 365432, 18 pp (*Preprint* `arXiv:1308.1633 [physics.ins-det]`)
[25] Xu W *et al.* 2015 The MAJORANA DEMONSTRATOR: A search for neutrinoless double-beta decay of ^{76}Ge *Proceedings for this workshop* (*Preprint* `arXiv:1501.03089 [nucl-ex]`)
[26] Szczerbinska B *et al.* 2010 *Proceedings of the South Dakota Academy of Science* **89** 23
[27] Berkeley low background facility `http://lbf.lbl.gov`
[28] Best A, Görres J, Junker M, Kratz K -L, Long A, Smith K and Wiescher M Low energy neutron background in deep underground laboratories To be submitted to *Astropart. Phys.* in 2015
[29] Tiedt D 2013 *Radioactive background simulation and cleanliness standards analysis for the Long Baseline Neutrino Experiment located at the Sanford Underground Research Facility* Master's thesis (unpublished) South Dakota School of Mines and Technology Department of Physics
[30] Harms J *et al.* 2010 *Class. Quant. Grav.* **27** 225011 (*Preprint* `arXiv:1006.0678 [gr-qc]`)
[31] Acernese F, De Rosa R, De Salvo R, Garufi F, Giordano G, Harms J, Mandic V, Sajeva A, Trancynger T and Barone F 2010 *J. Phys.: Conf. Ser* **228** 012036
[32] Gage J R, Fratta D, Turner A L, MacLaughlin M M and Wang H F 2013 *Int. J. Rock Mech. Mining Sci.* **61** 244–255
[33] Volk J *et al.* 2012 *J. Inst.* **7** P01004 (*Preprint* `arXiv:1205.1777 [physics.acc-ph]`)
[34] Volk J, Shiltsev V, Chupyra A, Kondaurov M, Singatulin S, Fratta D, Meulemans A, Potier C and Wang H 2012 *12th International Workshops on Accelerator Alignment IWAA, 10–14 September, Fermilab*
[35] Wang H F, Gage J R, Fratta D O, MacLaughlin M M and Turner A 2011 *Proceedings of the 8th International Workshop on Structural Health Monitoring, 13–15 September, Stanford, CA* 8 pp
[36] Gage J R, Noni N, Turner A, MacLaughlin M and Wang H F 2010 *44th U.S. Rock Mechanics Symposium and 5th U.S-Canada Rock Mechanics Symposium, 27–30 June, Salt Lake City, UT* 8 pp
[37] Kennedy J, Murdoch L, Long A J and Koth K 2011 *American Geophysical Union Fall Meeting, 5–9 December, San Francisco, CA* **Abstract NH54A-01** A1
[38] Steadman J 2015 *BIFs, black shales, and gold deposits: A reevaluation* Ph.D. thesis (unpublished) University of Tasmania School of Earth Sciences
[39] Armstrong A E 2013 *Hydrothermal alteration and gold mineralization in biotite zone host rocks at Homestake Mine in Lead, South Dakota* Master's thesis (unpublished) South Dakota School of Mines and Technology Department of Geology
[40] Hamer R C 2010 *Hydrothermal alteration and gold deposition in the Homestake iron-formation-hosted gold deposit, Lead, South Dakota* Master's thesis (unpublished) South Dakota School of Mines and Technology Department of Geology
[41] Roggenthen W M and Koch C D 2013 *Proc. 47th U.S. Rock Mech./Geomech. Symp. 23–26 June, 2013, San Francisco, CA* **Paper 13-493** 8
[42] Sherman C S, Magliocco M and Glaser S D 2011 *45th U.S. Rock Mech./Geomech. Symp., 26–29 June, San Francisco, CA* **Paper 11-526** 7
[43] Van Beek J K, Roggenthen W M, Magliocco M and Glaser S D 2009 *Proceedings of the 3rd CANUS Rock*

Mechanics Symposium, May 2009, Toronto, ON **Paper 4155** 7
[44] Rinke C *et al.* 2013 *Nature* **499** 431–437
[45] Bergmann D J, Anderson C M and Gorbatenko O 2014 *Proceedings of the South Dakota Academy of Science* **93** 214
[46] Finch A E, Brakke A, Gorbatenko O, Anderson C M and Bergmann D J 2013 *Proceedings of the South Dakota Academy of Science* **92** 173
[47] Waddell E J, Elliott T J, Vahrenkamp J M, Roggenthen W M, Sani R K, Anderson C M and Bang S S 2010 *Env. Tech.* **31:8–9** 979–991
[48] Bhalla A, Kainth A and Sani R K 2013 *Genome Announc.* **1** e00595–13
[49] Rastogi G, Gurram R N, Bhalla A, Jaswal R, Gonzalez R, Bischoff K M, Hughes S R, Sudhir K and Sani R K 2013 *Journal of Frontiers in Microbiotechnology, Ecotoxicology and Bioremediation* **4** 1–8
[50] Bhalla A, Bansal N, Kumar S and Sani R K 2013 *Bio. Tech.* **128** 751–759
[51] Rastogi G, Gurram R N, Bhalla A, Gonzalez R, Bischoff K M, Hughes S R, Kumar S and Sani R K 2013 *Front. Microbiol.* **4** 18
[52] Rastogi G, Bhalla A, Adhikari A, Bischoff K M, Hughes S R, Christopher L P and Sani R K 2010 *Bio. Tech.* **101** 8798–8806
[53] Rastogi G, Osman S, Kukkadapu R K, Engelhard M, Vaishampayan P A, Andersen G L and Sani R K 2010 *Microb. Ecol.* **60** 539–550
[54] Rastogi G, Muppidi G L, Gurram R N, Adhikari A, Bischoff K M, Hughes S R, Apel W A, Bang S S, Dixon D J and Sani R K 2009 *J. Ind. Microbiol. Biotechnol.* **36** 585–598
[55] Rastogi G, Stetler L D, Peyton B M and Sani R K 2009 *J. Microb.* **47** 371–384
[56] Murthy N K S, Bleakley B H and Speidel R 2011 *Proc. 61st Society for Industrial Microbiology Annual Meeting, 24–28 July, New Orleans, LA* **Abstract P99**
[57] Harris S T, Gibson K A and Bleakley B H 2010 *Proc. 60th Society for Industrial Microbiology Annual Meeting, 1–5 August, San Francisco, CA* **Abstract P35**
[58] Shende A, Harder-Heinze D and Shende R 2012 *Proceedings of the South Dakota Academy of Science* **91** 208
[59] Shende A, Jaswal R, Harder-Heinze D, Menan A and Shende R 2012 *Technical Proceedings of the 2012 NSTI Nanotechnology Conference and Expo* **3** 489
[60] Shende A, Jaswal R and Shende R 2012 *Proceedings of AIChE Annual Meeting* **ISBN: 978-1-4665-6277-6** 120
[61] Osburn M R, LaRowe D E, Momper L M and Amend J P 2014 *Front. Microbiol.* **5** 610:1–14
[62] SURF science liaison office `http://www.sanfordlab.org/science-liaison-office`
[63] 2014 Report of the particle physics project prioritization panel (p5) `http://www.usparticlephysics.org/p5`
[64] Best A *et al.* 2013 *Phys. Rev. C* **87** 045805 (*Preprint* `arXiv:1304.6443 [nucl-ex]`)
[65] Best A, Couder M, Famiano M, Lemut A and Wiescher M 2013 *Nucl. Instrum. Meth. A* **727** 104–108 `http://www.jinaweb.org/underground/DIANA`
[66] Adams C *et al.* 2013 Scientific opportunities with the Long-Baseline Neutrino Experiment (*Preprint* `arXiv:1307.7335 [hep-ex]`)
[67] Barker D, Mei D -M and Zhang C 2012 *Phys. Rev. D* **86** 054001 (*Preprint* `arXiv:1202.5000 [physics.ins-det]`)
[68] Agarwalla S K *et al.* 2014 *J. High Energy Phys.* **1405** 094 (*Preprint* `arXiv:1312.6520 [hep-ph]`)
[69] An experimental program in neutrinos, nucleon decay and astroparticle physics enabled by the Fermilab Long-Baseline Neutrino Facility `https://indico.fnal.gov/getFile.py/access?resId=0&materialId=0&confId=9214` Letter of Intent submitted to the Fermilab PAC

2nd Workshop on Germanium Detectors and Technologies IOP Publishing
Journal of Physics: Conference Series **606** (2015) 012016 doi:10.1088/1742-6596/606/1/012016

The GRETINA Spectrometer

M. Cromaz

Lawrence Berkeley National Laboratory, 1 Cyclotron Rd, Berkeley CA 94720
E-mail: mcromaz@lbl.gov

Abstract. The GRETINA spectrometer is a first generation, gamma-ray tracking spectrometer capable of determining the Compton scattering path of gamma-rays incident on the detector volume. This ability allows the Ge detectors to be close packed allowing the detector to be scaled to high efficiencies while maintaining good peak-to-total. GRETINA currently consists of 7 4-detector modules giving approximately 1π solid angle coverage with a calorimetric efficiency of 6.3% and tracked efficiency of 4.7% at 1.3 MeV. The array's sensitivity to the position of the gamma ray's first interaction point enables precision event-by-event Doppler correction which allows one to achieve 1% energy resolution even for sources moving at a large fraction of the speed of light such as those encountered at fragmentation facilities such as NSCL and the future FRIB.

1. Introduction

The availability of large volume, segmented HPGe crystals has enabled the development of the first generation of high-efficiency, general purpose, gamma-ray tracking spectrometers. Two such projects are currently being pursued: the Gamma-Ray Energy Tracking In-beam Nuclear Array (GRETINA) [1] in the United States and the Advanced GAmma Tracking Array (AGATA) [2] in Europe. The idea behind these spectrometers is to construct a spherical shell of Ge detectors around the target position of maximal efficiency. Interaction points of the incident gamma rays in the crystals are located, grouped into clusters, and tracked using the Compton scattering relation [3]. There are several distinct advantages to this approach compared to previous high-efficiency spectrometers such as Gammasphere [4]. Tracks with a set of interaction points which do not fit the Compton scattering relation, whose origin is assumed to be the target position, can be rejected as likely not depositing their full energy in the active detector volume of the array. This allows such arrays to operate with high peak-to-totals without the need for Compton suppression enabling the array to be scaled to high efficiencies. Furthermore, the ability to track the gamma ray means that multiple gamma rays depositing energy in a given crystal can in principle be disentangled thereby avoiding summing.

GRETINA began its first major physics campaign at NSCL/MSU in conjunction with the S800 spectrometer in 2012 and the value of this new technology was quickly established. The ability to apply precision Doppler correction of the gamma-ray energy based on the location of the first interaction point in the crystal enables one to maintain 1% energy resolution for sources moving at a high fraction (40%) the speed of light while maintaining high efficiency and peak-to-total. This allows for precision spectroscopy of isotopes close to the driplines produced in fragmentation reactions. GRETINA is now completing its second physics campaign at ATLAS/Argonne to take advantage of radioactive beams produced by Cf fission source CARIBU. In this paper we will give a

Journal of Physics: Conference Series **606** (2015) 012016 doi:10.1088/1742-6596/606/1/012016

brief description of the technical implementation of GRETINA, its online data analysis algorithms to determine the gamma-ray tracks, its current performance characteristics, and the outlook for future development.

2. Technical Description

The technical aspects of Ge tracking detectors are more complex then that of previous arrays based on Compton suppression. Interaction positions in the crystal are located by examining the charge on each segment during charge collection with (primarily) the hit segment determining the radial position of the interaction while the image charges in adjoining segments locate the interaction position in ϕ and z. To achieve good position resolution the crystal itself is highly segmented and each segment pad is instrumented with high-speed digitizers to record the waveform. For GRETINA with 28 crystals, each with 36 segments, 1000 such channels are required.

2.1. *The Detector Module*

GRETINA employs large-volume, 36-way segment HPGe crystals with base diameter of 8 cm and length of 9 cm that are tapered to allow for spherical packing. The tapering is that of an irregular hexagon to allow efficient tiling of the spherical shell. The geometry chosen requires 120 hexagons to tile the complete sphere. GRETINA currently consists of 7, 4-detector modules yielding approximately 1π solid angle coverage. Each detector is individually encapsulated and four crystals are packaged in a single cryostat as shown in fig. 1.

The central contact preamplifier for each crystal utilizes cold Field Effect Transistors (FET) inside the cryostat to optimize resolution while the segments employ warm FETs outside the cryostat. This choice was made to simplify servicing the large number of channels. Each segment and the central contact have a separate preamplifier board that is mounted in a compartment directly behind the crystals in the detector module. The large heat load introduced by the 148 preamplifiers in the module necessitates the use of a liquid cooling system to maintain the preamplifier compartment at a reasonable temperature.

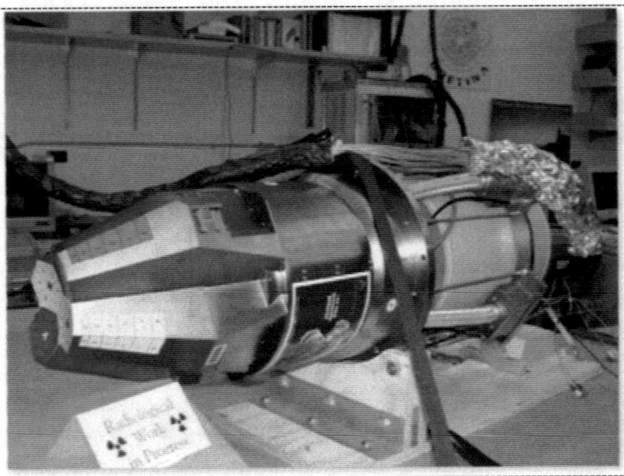

Figure 1. A GRETINA "quad" containing four hexagonally tapered HPGe crystals which can be assembled tile a spherical shell.

2.2 Electronics

In order to locate the interaction points within the segmented Ge crystal, it is necessary to record the charge on each segment pad as a function of time. This requires that each segment and the central contact to be instrumented with a high-speed digitizer to record these waveforms for analysis. To perform this function a specialized 10-channel, VME based digitizer was developed [5]. Banks of four such digitizers are used to instrument the 36 detector segments and the central contact. Signals are digitized at a frequency of 100 MHz with a dynamic range of 14 bits employing an AD645 ADC by Analog Devices. An onboard Field Programmable Gate Array (FPGA) (Xilinx XC3S5000) processes these data streams providing a leading edge discriminator used by the trigger systems, an energy, pileup flag and 48-bit timestamp which is used later in the processing pipeline for event building. The high-resolution energy is determined from the trace produced from the ADC by a trapezoidal filtering algorithm [6] which is baseline corrected.

GRETINA also has a programmable digital trigger system which initiates readout of the digitizers based on the number of prompt and/or delayed hits in the Ge central contacts, the trigger condition of external devices or combinations of the two. The trigger also distributes a global, 48-bit clock to all digitizers to synchronize the system to a 10ns clock and provide timestamps to uniquely assemble events from all the detector modules.

2.3. *Computing*

The computing system of GRETINA consists of several components. The local readout of digitizers is done using VME-based, single-board computers (Emerson MVME5500) with one such readout computer allocated for each crystal. These use the real-time operating system vxWorks upon which custom readout software was developed. This system is responsible for local event building at the crystal level, control of the digitizers, forwarding data to the online analysis system where the location of interaction points are determined, and basic performance monitoring.

The online data processing is done in a 63-node linux cluster. Each node has two four-core processors and 8Gb of memory. Data from the cluster is transferred from the local readout computers via a 10 Gb optical link utilizing a standard TCP-based network protocol. Most nodes in the cluster are allocated to the task of signal decomposition, which is the rate-limiting step, although resources are also be allocated for Compton tracking, global event building and monitoring. Data is stored to a network-based filesystem using the Network File System (NFS) protocol or to high-capacity storage nodes with local raid arrays.

As with Gammasphere, GRETINA employs a unified control system employing EPICS (Experimental Physics and Industrial Control System) [7]. This is a network-based, distributed control system that had been adapted to a wide-variety of computing platforms. In GRETINA it is used to configure the digitizers through the local readout computers, the trigger system as well as processes running on the cluster. This gives the experimenter a unified view of GRETINA as a whole. The system itself is also easily scriptable making it straightforward to develop configuration and diagnostic scripts for various detector configurations.

3. Signal Decomposition and Compton Tracking

In order to locate interaction points within the crystal, a procedure termed signal decomposition is performed [8]. Given the crystal geometry, impurity concentration and profile, one first simulates the signals on each segment contact for unit charge as a function of position in the crystal. This is done on a grid of points that span the volume of the detector and a library of such signals is created. Against this library linear combinations of basis signals are fit against the experimentally determined waveforms to determine the positions and relative energies of interaction points. In this way the experimental positions of the interaction positions are "decomposed" and assigned positions within the crystal.

The basis must be corrected for the electronic response of the detector module and electronics. These response parameters include shaping introduced by the preamplifier, time offsets, as well as integral and differential crosstalk. Correcting for differential crosstalk is especially important as this is easily confused with the image charge signals that locate interaction points in phi and z within the crystal.
Given the large number of response parameters that need to be fit and the difficulty in measuring differential crosstalk directly, an automated method known as the superpulse method [8] was developed. Here, a ^{60}Co source is placed at a fixed distance from the detector and the waveforms for events where a single hit segment are recorded. The same configuration is then simulated and the parameters are fit against the averaged experimental signals given a realistic response model. These parameters are then applied to each signal in the basis with the same response model. Detailed comparisons of fit response parameters with direct measurements of shaping time, differential and integrated crosstalk show this method to work well. An example is shown in Fig. 2 where integral crosstalk coefficients between segments of a crystal derived from both segment sums and the superpulse method are shown to be in good agreement.

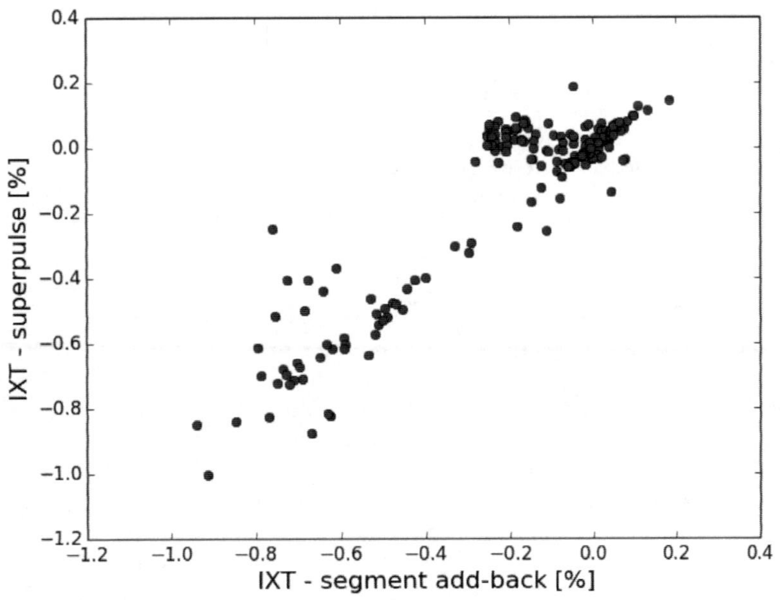

Figure 2. Comparison of integral crosstalk parameters (IXT) derived from the two-fold segment sum method (x-axis) and the those obtained through a superpulse fit (y-axis).

There are two main phases to the signal decomposition algorithm employed. The first is an adaptive grid search where a best fit to pairs of interaction points within a given segment are made on the grid points as defined in the basis. Following this a non-linear fit is performed where one can move off of grid points and the time offset is fit. Points that fall within a certain minimum distance of each other are coalesced into a single point weighted by their energies. This algorithm is computationally efficient requiring ~ 10ms/core on modern computing platform. The first hit position resolution (σ) of the interaction points has been determined to be 2 mm both from collimated source tests and inferred from in-beam Doppler correction. Fig. 3 shows a representative distribution of an interaction points from a collimated ^{137}Cs source.

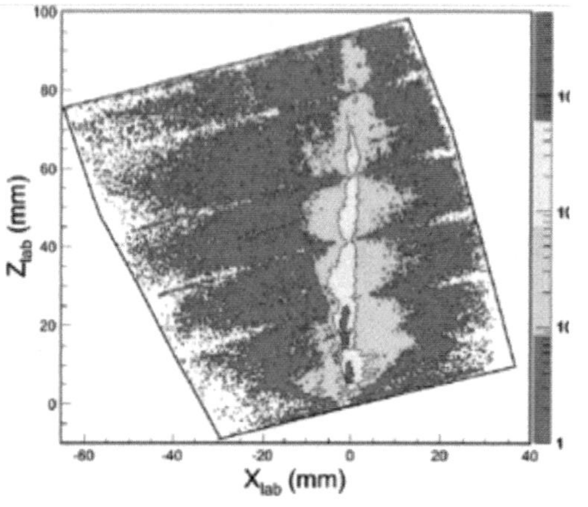

Figure 3. Distribution of interaction points from a ^{137}Cs source collimated with a 1 mm aperture incident on a GRETINA segmented HPGe crystal.

Following signal decomposition the interaction points from individual crystal are tracked [9]. Interaction points identified in the signal decomposition are time-correlated and sorted into global events and the position of the interaction points are transformed from local crystal coordinates to world coordinates. Interaction points are then clustered into groups of likely gamma rays based on their angular separation. Then group-by-group, one examines each permutation of points to find which best fit the Compton scattering relation. The best fit is then assigned a figure of merit value on which it is decided if the gamma-ray likely deposited its full energy in the array or scattered out.

4. **Performance**

There are several performance metrics that contribute to the overall sensitivity of the detector array which include energy resolution, efficiency, peak-to-total, and rate. These have been carefully studied during the physics campaigns at NSCL/MSU and at ATLAS/Argonne. The energy resolution of the central contact signals was found to be 2.4 keV at 1.3 MeV and 1.7 keV at 0.12 MeV when mounted in the array. The mean segment energy resolution is 2.6 keV at 1.4 MeV and 1.6 keV at 0.12 MeV. These measurements were taken with trapezoidal filter parameters of 5 μs for the integration time and 1 μs for flat-top time.

Journal of Physics: Conference Series **606** (2015) 012016 doi:10.1088/1742-6596/606/1/012016

As GRETINA employs tracking, the topic of efficiency is multifaceted as efficiency gains are realized by tracking gamma rays across multiple crystals. Hence, the specific geometric arrangement of the 7 modules in the frame is important. Also, as the location of interaction points given by signal decomposition is not exact, there is a tradeoff between efficiency and peak-to-total. The base calorimetric efficiency of the array with 7 modules in a close-packed configuration is 6.3%. This is essentially the arrays' maximum singles when tracking is not applied. As mentioned previously, if one applies a figure-of-merit cut to the fit of the Compton track of the gamma ray, one can significantly increase peak-to-total at the cost of efficiency. Defining a 25% FOM cut, which reduces the array efficiency to 6.3%(0.75) = 4.7%, one achieves peak-to-total of 49% (again in the close-packed configuration).

As with efficiency, there are also two aspects to the rate performance of GRETINA. The singles rate, the maximum rate at which a single Ge crystal can operate, is 20 kHz and is limited by the shaping time in the trapezoidal filter to retain good energy resolution without incurring excessive pileup. Unlike conventional spectrometers, GRETINA performs the additional step of online signal decomposition to locate interaction points. This occurs post-trigger and is limited by the computational power available for signal decomposition in the online signal processing cluster and network/backplane bandwidth constraints. The current maximum processing rate for signal decomposition is 30 kHz that, assuming an average of 1.5 detectors hit per gamma, yields a rate of 20000 gamma/s for the 7-module array post-trigger. This rate can be in principle increased by increasing processing power in the cluster and improving bandwidth from the digitizers to the readout controllers.

5. Outlook and Conclusion

The original GRETINA project specified an array of 28 crystals grouped in 7 modules to give 1π solid angle coverage. Since project completion two additional quad modules have been procured and a third is in process. The ultimate goal is to construct a shell with 4π solid angle coverage (30 modules). A second area of development is in signal decomposition and tracking algorithms. While the position resolution of the primary (generally first) hit is high, $\sigma = 2$ mm, the position resolution of secondary interaction points is degraded and is subject to producing false artifacts. A detailed, source-based, detector scanning program will be carried out to examine the response of the detector in detail and compare against simulation.

GRETINA has demonstrated the viability of gamma-ray tracking in large volume, coaxial Ge detectors which will allow such spectrometers to be built to unprecedented efficiencies and therefore sensitivities. Furthermore, its precision event-by-event Doppler correction capability due to its position sensitivity allows the spectrometer to maintain high energy resolution for high velocity sources such as those produced in fragmentation reactions. These capabilities have extended the science reach of current facilities such as NSCL and will prove key in carrying out the proposed science program of FRIB [10].

6. Acknowledgements

I would like to gratefully acknowledge the work of the GRETINA collaboration and in particular D. Weisshaar, D.C. Radford, A.O. Macchiavelli and I.Y. Lee. This work was supported by the Office of Science in the Department of Energy under grant DE-AC02-05CH11231.

7. References

[1] Paschalis S, Lee I Y, Macchiavelli A O, Campbell C M, Cromaz M, Gros S, Pavan J, Qian J, Clark R M, Crawford H L, Doering D, Fallon P, Lionberger C, Loew T, Petri M, Stezelberger T, Zimmermann S, Radford D C, Lagergren K, Weisshaar D, Winkler D, Glasmacher T, Anderson J T and Beausang C W 2013 *Nuclear Instruments and Methods in Physics Research A* **709** 44
[2] Simpson J 2005 *Acta Physica Polonica* B **36** 1383
[3] Deleplanque M A, Lee I Y, Vetter K, Schmid G J, Stephens F S, Clark R M, Diamond R M, Fallon P and Macchiavelli A O 1999 *Nuclear Instruments and Methods in Physics Research A* **430** 292
[4] Lee I Y, 1990 *Nuclear Physics A* **520** 641
[5] Anderson J, Brito R, Doering D, Hayden T, Holmes B, Joseph J Yaver H and Zimmermann S 2009 *IEEE Transactions on Nuclear Science* **56** 258
[6] Jordanov V T and Knoll G F 1994 *Nuclear Instruments and Methods in Physics Research A* **345** 337
[7] EPICS http://www.aps.anl.gov/epics/, 2015
[8] Radford D.C. to be published
[9] Schmid G, Deleplanque M A, Lee I Y, Stephens F S, Vetter K, Clark R M, Diamond R M, Fallon P, Macchiavelli A O and MacLeod R W 1999 *Nuclear Instruments and Methods in Physics Research A* **430** 69
[10] FRIB http://www.frib.msu.edu/, 2015

The AGATA Spectrometer: next generation gamma-ray spectroscopy

J Simpson[1]
STFC Daresbury Laboratory, Daresbury, Warrington, WA44AD, UK
E-mail: john.simpson@stfc.ac.uk

On behalf of the AGATA collaboration

Abstract. The Advanced GAmma Tracking Array (AGATA) is a European project to develop and operate the next generation gamma-ray spectrometer. AGATA is based on the technique of gamma-ray energy tracking in electrically segmented high-purity germanium crystals. The spectrometer will have an unparalleled level of detection power for electromagnetic nuclear radiation. The tracking technique requires the accurate determination of the energy, time and position of every interaction as a gamma ray deposits its energy within the detector volume. Reconstruction of the full interaction path results in a detector with very high efficiency and excellent spectral response. The realisation of gamma-ray tracking and AGATA is a result of many technical advances and the spectrometer is now operational. AGATA has been operated in a series of scientific campaigns at Legnaro National Laboratory in Italy and GSI in Germany and is presently being assembled at GANIL in France. The status of the instrument will be reviewed.

1. Introduction

High precision gamma-ray spectroscopy is one of the powerful tools used to study the structure of excited nuclear states and it has contributed to the discovery of a wide range of new phenomena. Each major advance in gamma-ray detection has resulted in significant new insights into the structure of nuclei. These advances culminated in two state-of-the-art 4π arrays of escape-suppressed spectrometers, Euroball in Europe [1] and Gammasphere in the USA [2]. These arrays, of more than a 100 large volume germanium crystals each surrounded by a Compton-suppression shield, have pushed this particular detector technology to its limit, with an efficiency for a 1 MeV gamma ray of about 10% and a peak-to-total ratio of 0.5.

Over the last decade or so the consensus of opinion worldwide is that the next major step in gamma-ray spectroscopy involves abandoning the concept of a physical suppression shield and achieving the ultimate goal of a 4π Germanium detector ball through the technique of gamma-ray energy tracking in highly segmented Ge crystals. Such an instrument will have a huge increase in sensitivity for the detection of the weakest signal from the most exotic nuclei. It will have an unprecedented angular resolution, resulting in excellent energy resolution by minimizing Doppler broadening even at velocities of the emitting nuclei up to 50% of the velocity of light. It is superbly suited to be used in conjunction with the new generation of radioactive beam accelerators or existing stable beam facilities. It will be vital to realize the scientific potential of new facilities for example at

[1] john.simpson@stfc.ac.uk

FAIR, FRIB, HIE-ISOLDE, SPES, and SPIRAL2. Given the importance of this increase in sensitivity, collaborations have been established in Europe and the USA to construct 4π tracking spectrometers. In the USA the project is called GRETA (Gamma Ray Energy Tracking Array) [3] and in Europe AGATA (Advanced Gamma Tracking Array) [4]. Both these projects have taken the first step with the construction of GRETINA (Gamma Ray Energy Tracking In beam Nuclear Array) and first phases of AGATA. These projects represent major technological advances in radiation detection and each has taken a very similar overall approach. This article summarises very briefly the status of AGATA.

2. The elements of gamma-ray tracking

The improved sensitivity or resolving power is due to the new technique of tracking, which identifies the position and energy of gamma-ray interaction points in the detector segments. Since most gamma rays interact more than once within the crystal, the energy-angle relationship of the Compton scattering formula is used to track the path of a given gamma ray. The full gamma-ray energy is obtained by summing only the interactions belonging to that particular gamma ray. In this way there are no lost scatters into suppression shields and scattered gamma rays between crystals are recovered, thus a much higher overall efficiency can be achieved. A gain by a factor of 6 over Gammasphere/Euroball for a single 1 MeV gamma ray and about 20 times for 15 MeV gamma rays is expected. Other key benefits of a highly segmented Ge array include high energy resolution, high counting rate capability (a factor of ~5 over Gammasphere/Euroball), good angular position resolution (about 1° versus 6° for Gammasphere) critical for Doppler shift corrections since many experiments involve high recoil velocities, the ability to handle high multiplicities without summing (as interactions from different gamma rays can be resolved), and the ability to pick out low-multiplicity events hidden in a high background environment due to the possibility of rejecting background events using the directional information of the radiation. The segmentation plus tracking also makes high precision linear polarisation measurements possible. The modularity of the detector design makes it extremely versatile and flexible for use in many different configurations and in conjunction with many ancillary detectors.

The realisation of such a system has required the development of highly segmented germanium detectors, digital electronics, pulse shape analysis to extract energy, time and position information and tracking algorithms to reconstruct the full interaction.

The detectors of AGATA (and GRETINA) are highly segmented n-type coaxial segmented germanium detectors supplied by the company Canberra, France. The crystals have a length of 9 cm and an initial diameter of 8 cm. Each crystal is shaped to form a hexagonal geometry and tapered at the front at an 8^0 angle. The tapering arises from the design of the full 4π detector ball being based on the geodesic arrangement of irregular 180 hexagons and 12 regular pentagons. The geometry and performance of AGATA is discussed in [5]. Each crystal is encapsulated into a thin Al can to protect the surface of the crystal. The crystals are segmented on their outer surface to provide the first level position of interaction information. The segmentation is 6 fold both longitudinally and azimuthally. This results in 37 signal outputs from the detectors, 36 segments and 1 signal from the core. The performance of the individual detectors is excellent meeting the specifications, for example with the core energy resolution being less than 2.35 keV at 1.33MeV and the segments less than 2.3 keV. In the AGATA array there are three slightly different hexagonal shapes, which can be grouped easily in the geometry. Three detectors are therefore assembled together and housed in a so-called triple cryostat. This cryostat [6], which is provided by the company CTT Montabaur, cools the detectors and houses the 111 cold FETS and collaboration designed preamplifiers. Figure 1 shows a CAD design of the spectrometer and its realisation with 5 triple clusters (15 Germanium detectors) mounted for the experimental campaign at Legnaro National Laboratory (LNL).

The segmentation of the Ge crystal provides the initial position-of-interaction information. In order to perform gamma-ray tracking, the positions and energies of the gamma-ray interactions in the Ge must be even more accurately determined from the signal waveforms using pulse shape analysis. The position resolution is a key metric in the detector performance since it directly affects the efficiency and peak/total of the array, as well as the effective energy resolution of the array when the source of radiation is travelling with high recoil velocity. AGATA requires a state-of-the-art, purpose-built digital electronics and an associated data acquisition system to process the signals from the Ge crystals to enable accurate position information to be obtained. The system comprises an initial digitiser to sample each of the 37 signals to timestamp each signal (to 10ns), provide the energy of each segment and core and a digitised sample of the leading edge of each pulse. The latter is used later in the pulse shape analysis to determine the position of interaction. Following further processing the system collates and distributes the events to a computer cluster to perform pulse shape analysis, event reconstruction and tracking. A Global clock and trigger system has been developed to synchronise and time align the signals. The system needs to cope with a large numbers of channels, over 6000, and also with very high rates in each crystal, up to 50 kHz. Interfaces to enable coupling of ancillary detectors to the acquisition system have been developed. In the first phase of the AGATA project a full electronics and data acquisition system was provided for up to 30 detectors.

Crucial to the successful operation of a tracking array is the extraction of the position of each interaction within an event. The position and energy of a given interaction in the detector is found using a procedure called signal decomposition (pulse shape analysis). A computed signal basis is pre-calculated for each crystal using a simulation where a unit charge is placed at a given point in the crystal and the net and transient currents induced on each of the 36 segment contacts are calculated. Measured signals are then compared with linear combinations of these simulated signals, as there are typically multiple interaction points in a given crystal, with the best fit giving the location and charge (energy) of the interaction points. The calculation involves the geometry of the crystal, the bias voltage, and a model of the space charge distribution arising from crystal impurities. This idealised model of the crystal is not sufficient when applied to real signals, since corrections must be made for detector and electronic-response characteristics such as preamplifier shaping, relative time delays, and integral and differential cross-talk. These corrections are determined by comparing signal traces measured using gamma-ray sources to the signal basis, and fitting the required parameters. AGATA has a programme of scanning and characterising the crystals to provide a database of signals needed for validation of the simulations, for example see [7].

It has been shown experimentally that the pulse shape algorithms developed for AGATA can achieve an average position resolution of better than 5 mm (FWHM), which is sufficient for efficient gamma-ray tracking.

3. The deployment of AGATA

The AGATA collaboration of over 40 institutions in 13 countries in Europe was established to develop and construct a 4π tracking spectrometer. The AGATA collaboration defined that the full array would be realized in phases. In 2003 a Memorandum of Understanding (MoU) was signed by the partners for the research and development phase of the project. This first phase of the project involved constructing an array of 15 detector capsules, with all the electronics and data acquisition, pulse shape and tracking algorithms and associated spectrometer infrastructure. This phase of the project is complete and AGATA has commenced a series of science campaigns.

AGATA is an instrument that moves between laboratories in order to take full advantage of the different beams and facilities available, and hence to maximise the breath of science that is addressed. In the first phase, AGATA was located at Legnaro National Laboratory up to the end of 2011[8]. AGATA was coupled to the PRISMA magnetic spectrometer and also used with a plunger for lifetime measurements and a suite of additional detectors, see [8]. A series of in-beam commissioning tests was performed followed by a successful science campaign of 23 experiments.

The spectrometer was then moved and operated at the GSI laboratory in Germany. At GSI AGATA was used to perform spectroscopy using the very high energy secondary beams from the UNILAC/SIS accelerator complex. At GSI the secondary beams of radioactive nuclei are produced by fragmentation of a range of stable primary beams and/or fission of a ^{238}U beam on a ^{9}Be or ^{208}Pb target placed at the entrance of the fragment recoil separator (FRS). The secondary beams were transported to the focal plane of the FRS where gamma-ray spectroscopy is performed using techniques such as relativistic Coulomb excitation and secondary fragmentation. The secondary beams are tracked through the FRS, and following reactions on a secondary target at the focal plane, are identified in terms of A and Z. For the GSI campaign AGATA consisted of up to 25 detectors. The reactions used at GSI result in high velocity recoils and a forward focusing of the gamma radiation emitted. Following extensive simulations and taking into account the mechanical design constraints the optimum efficiency was calculated to be achieved with only two detectors each cryostat in the most forward direction. A new double cryostat was therefore developed and the array was designed for 5 double and 5 triple cryostats, 25 detectors.

AGATA ceased operation at GSI in March 2014 and is currently being set-up and commissioned at the GANIL laboratory in France. The GANIL facility offers a large variety of radioactive and stable beams, ranging from ^{6,8}He to ^{238}U, over a wide range of energies from a few MeV/u up to 100 MeV/u. This wide spectrum of beams combined and the availability of several existing state-of-the-art spectrometers for gamma-ray and particle detection, such as VAMOS, has resulted in the collaboration deciding to host AGATA at GANIL until the end of 2018. At this time the instrument will have in excess of 45 detectors. Following this GANIL phase the instrument will return to LNL to take advantage of the radioactive beams from new facility, SPES.

4. Science advances, Summary and outlook

The advances made in Ge detector technology, digital data acquisition systems, signal decomposition and gamma-ray interaction reconstruction have proved that gamma-ray tracking spectrometers can be successfully deployed. AGATA and GRETINA (and GRETA) will have an enormous impact on the exploration of nuclear structure at the extremes of isospin, proton number, angular momentum, excitation energy, and temperature. It is encouraging that the first scientific results from AGATA are now regularly appearing in scientific journals that are advancing our knowledge of the nuclear landscape. An excellent example of the power of AGATA used in conjunction with PRISMA is the spectroscopy of the neutron rich nucleus ^{196}Os by John et al., [9]. The scientific papers are the result of time consuming hard work in analysing this new rather and complex data. AGATA is now proving to be able to make a major advance in gamma-ray detection sensitivity that will enable the discovery of new phenomena, which are only populated in a tiny fraction of the total reaction cross section or in nuclei that are only produced with rates of the order of a few per second or less. The unprecedented angular resolution facilitates high-resolution spectroscopy with fast and ultrafast fragmentation beams giving access to the detailed structure of the most exotic nuclei that can be reached.

The instrumentation and technical advances driven by this work, and the knowledge gained by those involved, is important in a wide range of applications. These advances have potential impact in areas such as medical imaging in SPECT and PET systems, homeland security, and environmental monitoring.

The realisation of AGATA is the result of a tremendous amount of hard work of a huge number of people in many laboratories and the support of their funding agencies. I would like to take the opportunity to thank all those involved.

5. References

[1] Simpson J 1997 *Z. Phys.* A **358** 139
[2] Delapanque M A and Diamond D 1987 Gammasphere Proposal, Preprint LBNL-5202
[3] Lee I Y, Deleplanque M A and Vetter K 2003 *Rep. Prog. Phys.* **66** 1095
[4] Akkoyun et al., 2012 *Nucl. Instrum. Meth. Phys. Rev.* A **668** 26
[5] Farnea E, Recchia F, Bazzacco D, Kroll Th, Podolzak Zs, Quintana B and Gadea A 2012 *Nucl. Instrum. Meth. Phys. Rev.* A **621** 331
[6] Wiens A, Hess , Birkenbach B, Bruyneel, Eberth, J, Lersch D, Pascovici G, Reiter P and Thomas H-G 2010 *Nucl. Instrum. Meth. Phys. Rev.* A **618** 223
[7] Colosimo S, et al., 2014 *Nucl. Instrum. Meth. Phys. Rev.* A **773** 124
[8] Gadea A et al., 2011 *Nucl. Instrum. Meth. Phys. Rev.* A **654** 88
[9] John P R, et al., 2014 *Phys. Rev.* C **90** 021301(R)

Germanium Detectors in Homeland Security at PNNL

S Stave[1]
Staff Scientist, Pacific Northwest National Laboratory,
Richland, WA 99352

E-mail: Sean.Stave@pnnl.gov

Abstract. Neutron and gamma-ray detection is used for non-proliferation and national security applications. While lower energy resolution detectors such as NaI(Tl) have their place, high purity germanium (HPGe) also has a role to play. A detection with HPGe is often a characterization due to the very high energy resolution. However, HPGe crystals remain small and expensive leaving arrays of smaller crystals as an excellent solution. PNNL has developed two similar HPGe arrays for two very different applications. One array, the Multisensor Aerial Radiation Survey (MARS) detector is a fieldable array that has been tested on trucks, boats, and helicopters. The CASCADES HPGe array is an array designed to assay samples in a low background environment. The history of HPGe arrays at PNNL and the development of MARS and CASCADES will be detailed in this paper along with some of the other applications of HPGe at PNNL.

1. Introduction

Neutron and gamma-ray detection is used for non-proliferation and national security applications. Non-proliferation examples include material identification and assay in safeguards applications or air monitoring in treaty verification. In the area of national security, neutron and gamma-ray detection is used in portal monitors to screen commercial and personal vehicles, aerial and ground radiation surveys (both of background radiation or by radioisotope), stand-off detection, and gamma-ray imaging. Pacific Northwest National Laboratory (PNNL) is involved in a variety of areas where high purity germanium (HPGe) presents an excellent solution. The work performed spans the space from commercial off-the-shelf systems to custom systems. In the area of fundamental science, PNNL has contributed to the MAJORANA neutrino-less double beta decay search [1], the CoGeNT detector for dark matter search [2], and assay of materials in a shallow underground lab for use in low background detectors[3]. In addition, there have been contributions to treaty monitoring through the RL-16 certified count lab, the Radionuclide Aerosol Sampler/Analyzer (RASA) [4], on-site inspection work utilizing commercial systems for surveys and research into information barriers.

The focus of this article will be on example applications that utilize HPGe arrays that have been developed at PNNL. It will be demonstrated that the requirements for HPGe arrays in the field are very different from those in a laboratory.

[1] 902 Battelle Blvd, PO Box 999, MSIN J4-65, Richland, WA 99352

2. High purity germanium

Sodium Iodide with thallium doping, NaI(Tl), is commonly used in field and laboratory gamma-ray detection applications due to its relatively low cost and high detection efficiency. Of course, NaI(Tl) scintillators typically have poor energy resolution relative to the gold standard HPGe. The use of HPGe allows the precise characterization of sources of gamma-ray radiation and oftentimes, a detection with HPGe is a characterization. **Figure 1** shows gamma-ray spectra collected with the MARS HPGe detector array (described below) and a 2”×4”×16” NaI(Tl) detector while observing the combination of a Pu and a ^{133}Ba source at 5 m. The weaker gamma-ray peaks from ^{239}Pu and ^{241}Am are very difficult to discern in the NaI(Tl) spectra due to the presence of the much stronger peaks associated with ^{133}Ba. However, the HPGe detector is able to easily make the distinction. Small, handheld germanium detectors are available and can be used but have efficiency limitations especially in comparison with available NaI(Tl)-based systems.

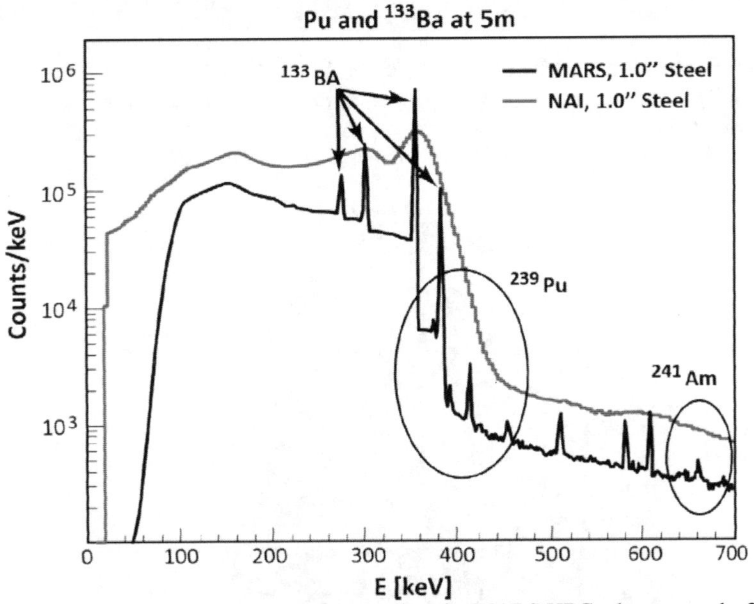

Figure 1. Example of a gamma-ray spectrum acquired using the MARS HPGe detector and a 2x4x16” NaI(Tl) detector exposed to a combined Pu and Ba-133 source at 5 m.

3. High purity germanium arrays

The obvious way to increase detection efficiency is to use larger detectors. HPGe crystals are limited in size by current production methods. The largest single crystal HPGe detectors are also the most expensive. A more cost effective solution is to combine multiple smaller crystals together into larger arrays thus balancing cost and size. The details of three such HPGe arrays will be discussed.

3.1. Previous Large HPGe Array

A previous large HPGe array was cooperatively developed by PNNL and Special Technologies Laboratory. It was fielded in the late 1980’s and demonstrated the use of broad-area flights over potentially radionuclide contaminated sites. The Environmental Radionuclide Sensor System (ERSS) [5] consisted of 20 60% relative efficiency HPGe crystals in two 10 crystal arrays. The system had a four hour mission time utilizing 50 liters of liquid nitrogen. The system had a 311 kg pod and three electronics racks for a total mass of 816 kg. Note that this total includes two ^{3}He-based neutron detectors and the supporting electronics. However, the majority of the mass was due to the HPGe system. The effects of vibration (specifically microphonics) were mitigated by inserting insulation into the cryostats. While this reduced the vibration effects, it also created numerous thermal shorts which contributed greatly to the heat load. While the system performed adequately, any new system required several improvements.

3.2. Multi-sensor Airborne Radiation Survey

The Multi-sensor Airborne Radiation Survey (MARS) detector [6] was designed using the lessons learned from the earlier ERSS array. MARS consists of 14 35% relative efficiency p-type HPGe crystals located in a single cryostat. Two platters of seven crystals each are tightly packed and minimize the material between crystals. The system was assembled and serviced in a PNNL cleanroom. Each HPGe crystal is individually biased and has its signal sent to an XIA Pixie-4 digitizer where the pulse height and time of the event are recorded. The signal from each crystal is then converted from pulse height to energy units and events within a short coincidence window (typically 1 microsecond) can be combined to improve the full energy photopeak efficiency. The cryostat and electronics have a combined weight of 140 kg in the helicopter mount. The cryogenic performance was improved by a factor of 100 with 46 hours between liquid nitrogen fills of 5 liters. The array attained an energy resolution of 3 keV at 1333 keV and the average energy resolution of each crystal was 2.9 at the same energy.

Figure 2. The MARS detector mounted in a side-looking configuration.

The MARS detector has had an active operational history. It has been transported across the United States multiple times to field testing in Las Vegas, NV, Charleston, SC, and Newport, RI along with a nuclear astrophysics measurement in Durham, NC and outdoor testing performed at PNNL's Building 3440 Test Track facility. In addition, it has been transported by air to an overseas maritime demonstration. Throughout all of the transportation, the system has continued to perform well even through the thermal cycle required for the air transport.

For the maritime testing on the US east coast and overseas, the system was operated on three different boats with similar system architectures. The electronics have an independent computer that is capable of acquiring the data and relaying back to land-based computers via wireless communications.

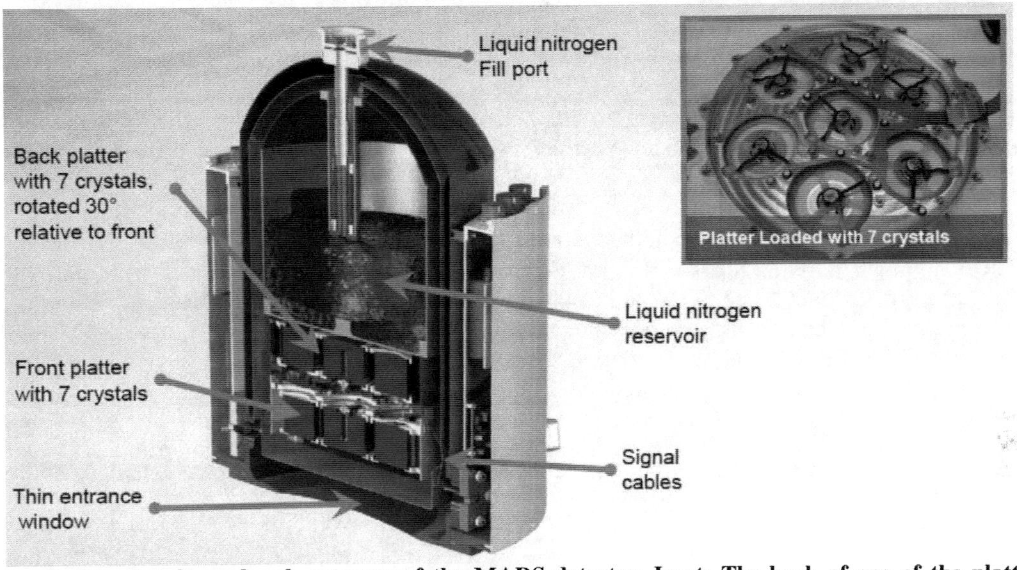

Figure 3. Annotated, rendered, cutaway of the MARS detector. Inset: The back of one of the platters holding seven crystals and showing custom flexible circuit board wiring.

3.2.1. Maritime Applications of MARS

Clearly, shielding and distance work against detector efficiency and identification. The solution is to get the detector closer to the potential sources and to do more with the detected gamma rays. Testing was performed with a variety of sources and with the detector being moved on a boat at different speeds, distances and orientations to the sources which were located on a larger, target vessel. Also, when developing hardware, it is critical to also develop software that is user friendly and is tailored to the target audience.

3.2.2. Land-based Testing of MARS

The maritime testing of the MARS detector presented a realistic environment but it was far from reproducible on demand as would be expected in a systematic detector characterization test. This systematic testing was performed on land in a controlled, static environment at PNNL's 3440 Test Track. The detector's response to different sources at different distances (5 to 100 m) and with different amounts of shielding was tested over the course of two weeks.

3.2.3. Aerial Testing of MARS

Returning to the original aerial application, the MARS detector mount design was modified so that it could be mounted in a helicopter with the detector facing downward rather than sideways. The task was to identify the advantages offered by a large HPGe array versus NaI(Tl) for aerial measurements. Five test flights were flown on and in the vicinity of the Nevada National Security Site (NNSS) near Las Vegas, NV:

- An initial checkout flight to identify and fix any early issues

Journal of Physics: Conference Series **606** (2015) 012018 doi:10.1088/1742-6596/606/1/012018

- An altitude spiral for determining the effects of air attenuation and cosmic ray background as a function of altitude over land and water
- A survey over part of the NNSS
- Flyovers of a variety of radioactive point sources multiple times at a sequence of altitudes
- Survey of a variable natural background region

Each flight served a particular testing purpose. Also, throughout all the flights, a three log RSI NaI(Tl) system was flown and acquired simultaneous data for comparison.

The clearest example of the advantage of HPGe over NaI in an aerial platform is shown in **Figure 4**. While the overall number of counts from the HPGe array is less than the NaI array, the specificity of the HPGe is clear. The HPGe array is able to do more with the fewer gamma-rays it detects due to the much better energy resolution. The individual peaks from ^{152}Eu are clearly evident in the HPGe plot and much harder to observe above the natural ^{40}K, ^{238}U, and ^{232}Th background in the NaI array.

The mapping ability of the MARS array was also able to be demonstrated with the natural background, gross counts map. The same features were visible in both the NaI and HPGe maps. Although an HPGe array would typically not be used in a gross counts mode, this test showed that synchronization of the GPS and MARS data streams was good.

The MARS array was also used to generate isotopic maps over the NNSS. While, an isotope with a single isolated gamma ray like ^{137}Cs is observable with both systems, it was already shown in **Figure 4** that a different isotope such as ^{152}Eu is much more readily mapped with an HPGe system.

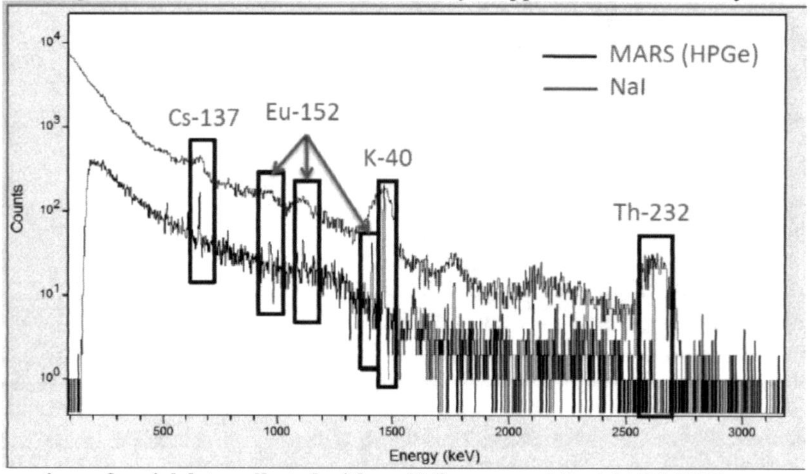

Figure 4. Comparison of aerial data collected with an NaI array and the MARS HPGe array.

3.3. CASCADES HPGe Array

While the MARS HPGe detector array was being developed, a similar array was also being constructed but for an entirely different application. The CASCADES detector [7] is also a 14 crystal array but the two seven crystal platter arrays are in independent cryostats and separated by enough space to allow the placement of samples between the two halves of the array. The entire array is in a lead cave surrounded by an anti-cosmic shield and is located in PNNL's Shallow Underground Lab. The Lab has approximately 30 meters water equivalent of overburden which is sufficient to decrease the fast neutron flux by a factor of 100 and the muons by a factor of 6.

Signals from each crystal in the CASCADES array are recorded with an XIA Pixie-4 digitizer system. This laboratory-based system has achieved 1.72 keV full width half maximum (FWHM) energy resolution at 662 keV and 2.13 keV FWHM at 1332 keV. Great attention was paid to the materials used to construct the array so as to minimize the background present in the detector itself. Recent improvements in the background performance of the array are shown in **Figure 5**. By lowering the intrinsic background of the detector, the minimum detectable activities of various isotopes are decreased. As an example, after the background improvements CASCADES has a minimum detectable activity of 6 mBq for Cs-137 in a 24 hour count. The system can provide useful results for very low activity samples as well as provide a high sensitivity capability to detect fission products.

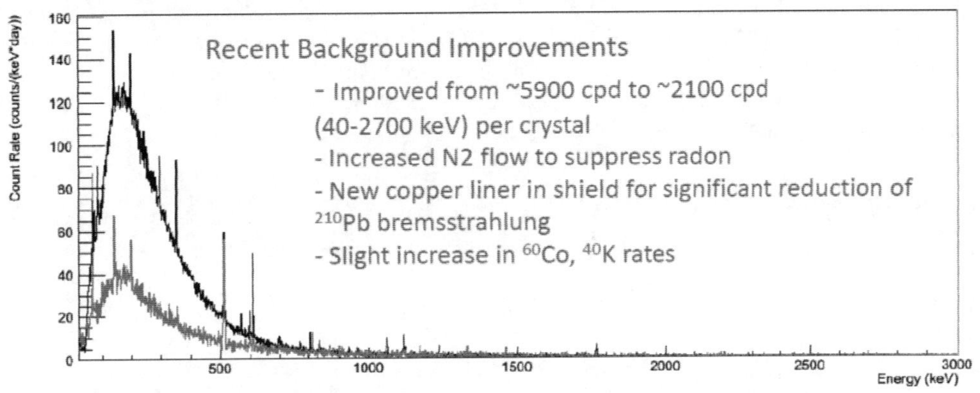

Figure 5. Example of the recent background improvements in the CASCADES array.

Through a high efficiency, high selectivity, and low background technical approach, the CASCADES array can be used in a variety of applications:

- Assay of atmospheric filter papers
- Resolve anomalous results seen in field measurements or normal laboratory assay
- Assay low level environmental samples
- Perform physics measurements related to gamma-ray branching ratios for particular isotopes
- Potential application to International Monitoring System samples

4. Conclusions

Two similar HPGe arrays developed at PNNL for two different applications have been detailed. Use of these arrays in the field and with a variety of samples has improved the understanding of these arrays. While useful, some need for improvements has been identified. For a fieldable array, the need for longer mission times and less logistics is desired. A move to a mechanical cooler would help alleviate the problems associated with procuring and using liquid nitrogen in the field as long as the vibrational effects of the cooler can be mitigated/tolerated.

High purity germanium detectors are already in the field in the form of small, portable detectors that are used for secondary inspections, as an example. However, as long as individual HPGe crystals remain relatively small and expensive, HPGe arrays will continue to have a place both in the field and in the laboratory.

Acknowledgments
The author would like to acknowledge the contributions from the PNNL MARS Flight Analysis Team: J.M. Benz, K.E. Dorow, J. Fast, Z.N. Gastelum, B.D. Glasgow, S. Morris, C.E. Seifert, J.M. Shergur, and J.A. Willett; and the PNNL CASCADES Team: C. Aalseth, A. Day, L. Erikson, B.D. Glasgow, E. Hoppe, T. Hossbach, L. Mizouni, M. Keillor, A. Myers, C. Overman, A. Seifert, and T. Stavenger. Portions of this work were funded by the Office of Defense Nuclear Nonproliferation Research and Development within the U.S. Department of Energy's National Nuclear Security Administration. PNNL is operated for the US Department of Energy by Battelle under contract DE-AC05-76RLO 1830. This is document number PNNL-SA-107425.

References

1. Aalseth, C.E., et al., *The Majorana Experiment.* Nuclear Physics B - Proceedings Supplements, 2011. **217**(1): p. 44-46.
2. Aalseth, C., et al., *CoGeNT: A search for low-mass dark matter using p-type point contact germanium detectors.* Physical Review D, 2013. **88**(1): p. 012002.
3. Aalseth, C.E., et al., *A shallow underground laboratory for low-background radiation measurements and materials development.* Review of Scientific Instruments, 2012. **83**: p. 113503.
4. Miley, H.S., et al., *A description of the DOE Radionuclide Aerosol Sampler/Analyzer for the Comprehensive Test Ban Treaty.* Journal of Radioanalytical and Nuclear Chemistry, 1998. **235**(1-2): p. 83-87.
5. Lepel, E., *"Field-deployable, aircraft-mounted sensor for the environmental survey of radionuclides".* Jour. Rad. Nucl. Chem., 1998. **112**: p. 211.
6. Fast, J.E., et al., *The multi-sensor airborne radiation survey (MARS) instrument.* Nucl. Inst. Meth. A, 2013. **698**: p. 152-167.
7. Keillor, M.E., et al., *Design and Construction of an Ultra-Low-Background 14 Crystal Germanium Array for High Efficiency and Coincidence Measurements.* Journal of Radioanalytical and Nuclear Chemistry, 2009. **282**(3): p. 703-708.

Study well-shaped germanium detectors for low-background counting

W.-Z. Wei[1], D.-M. Mei[1,2] and C. Zhang[1]

[1]Department of Physics, The University of South Dakota, Vermillion, SD 57069, United States
[2]School of Physics and Optoelectronic, Yangtze University, Jingzhou 434023, China

E-mail: `Dongming.Mei@usd.edu`

Abstract. Radiogenic particles are known as the main sources of background for all ultra-low background experiments in the detection of dark matter and neutrino properties. In particular, the radiogenic gamma rays from PMTs are a main component of the observed backgrounds in the noble liquid detectors such as XENON100 and LUX. This suggests a more accurate screening of PMTs is needed for the next generation experiments such as LUX-Zplin or Xenon1T. Hence, we propose to develop well-shaped germanium detectors for a better understanding of the radiogenic background from PMTs. A well-shaped germanium detector array and PMT (R11410MOD) have been designed in a Geant4-based Monte Carlo simulation, in which three radiogenic background isotopes from ^{238}U, ^{232}Th and ^{40}K have been studied. In this work, we show the detector performance including the detector efficiency, energy resolution and the detector sensitivity for low-background counting in the detection of rare event physics.

1. Introduction

Radiogenic background plays a critical role in a variety of ultra-low background experiments conducted deep underground for the detection of dark matter and neutrino properties, as radiogenic particles are known to be the main sources of background in those experiments. For example, radiogenic neutrons can cause isolated recoils of atomic nuclei that are indistinguishable from the expected signal of Weekly Interacting Massive Particles (WIMPs), the most promising candidate of dark matter.

Dark matter experiments using noble liquid xenon as detector material, such as XENON100 and LUX, typically employ photomultiplier tubes (PMTs) to detect the scintillation light from the xenon nuclear recoils [1]. PMTs are usually mounted directly above and below the active scintillation region for better detection of photon signals in the scintillation detector [1]. Hence, much attention needs to be paid on the radiogenic background generated from PMTs to maintain low background around the xenon detectors. This necessitates a good understanding of the radioactivity in PMTs, so that the PMTs selected for ultra-low background experiments will be comprised of radiopure materials [1]. In particular, the rediogenic gamma rays from PMTs are a main component of the observed backgrounds in XENON100 [2] and LUX [3], with an observed background rate often higher than what was predicted [2, 3]. This suggests a more accurate screening of PMTs is needed for the next generation experiments such as LUX-Zeplin [4] or Xeonon1T [5] in achieving the target sensitivity for dark matter searches.

Journal of Physics: Conference Series **606** (2015) 012019 doi:10.1088/1742-6596/606/1/012019

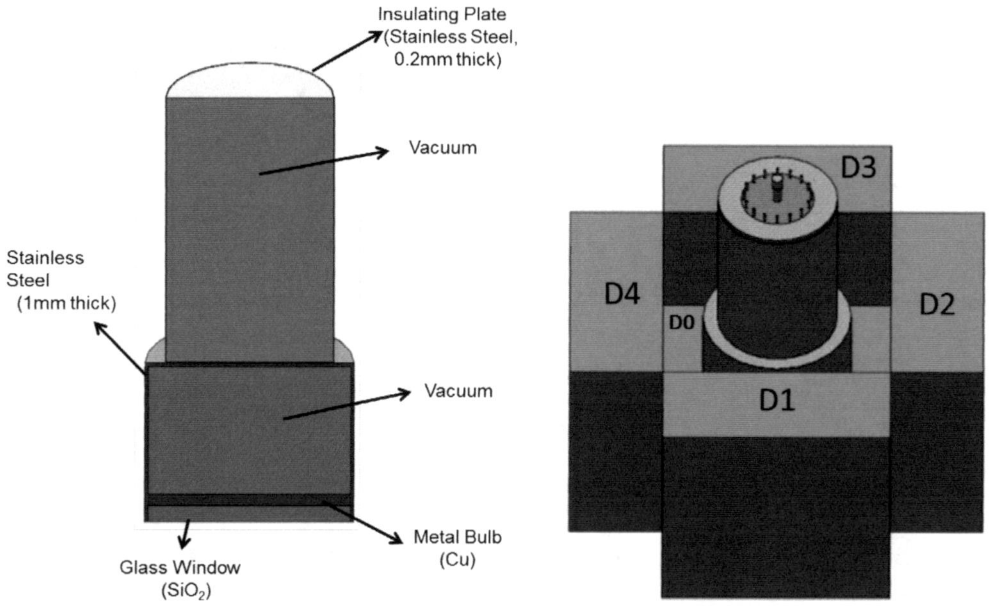

Figure 1. Simulated geometry for Ge well detector array and R11410MOD PMT. Left: Cross section of R11410MOD PMT without the lid. Right: Ge well detector array with the PMT in the center of the well, where D0 $\sim$ D4 represent the corresponding Ge planar detector.

2. Monte Carlo simulation set-up

2.1. Geometry and material

The proposed Ge well detector was constructed by making five Ge planar detectors in a well shape. Each Ge crystal is about 2.13 kg in mass with the dimension of 10 cm $\times$ 10 cm $\times$ 4 cm; and each single crystal is inside a very thin Oxygen-Free High-Conducting (OFHC) copper box with 2 mm in thickness. Furthermore, a R11410MOD PMT with dimensions and materials provided in Ref. [6] was simulated as well to provide the signal for the germanium well detector array in order to test how well this detector array will see the signals from PMT. Fig. 1 shows the geometry simulated in the Monte Carlo.

2.2. Generator

Three radiogenic isotopes, ^{238}U, ^{232}Th and ^{40}K, have been generated in the simulation. These three isotopes are prominent contaminants in the PMT and OFHC copper and they are all radioactive and will emit alpha and beta particles in their decay chains. To study the detector performance mentioned in the introduction, we have designed two independent simulations:

(I). Generate ^{238}U, ^{232}Th and ^{40}K only in the PMT to study the detector response to the signal from the PMT. ^{238}U, ^{232}Th and ^{40}K are distributed uniformly in each component material of the PMT which consists of copper, stainless steel, glass and ceramic. The simulated decays of each isotope in different material of the PMT is proportional to the mass of material, as shown in table 1. Four independent sub-simulations corresponding to four component materials have been conducted for each isotope and then combine them together to study the detector efficiency for a series of gamma rays of interest from those three sources. The detector efficiency will show how well the detector will respond to the signal.

(II). Generate ^{238}U, ^{232}Th and ^{40}K only in the five copper boxes to study the detector

Table 1. The simulated radioactive decays of each source in different materials of PMT. The total mass of PMT is ~202 g. 10^7 is the total decays simulated for each source.

Material	Mass (g)	Mass Percentage (%)	Simulated Decays
Copper	96.18	47.74	$10^7 \times 47.74\% = 4.774 \times 10^6$
Stainless Steel	63.23	30.89	$10^7 \times 30.89\% = 3.089 \times 10^6$
Glass	35	17.37	$10^7 \times 17.37\% = 1.737 \times 10^6$
Ceramic	8.06	4	$10^7 \times 4\% = 4.0 \times 10^5$

response to the background from the shield, copper box. ^{238}U, ^{232}Th and ^{40}K are uniformly distributed in the copper box. To obtained how many decays should be simulated for each isotope in the OFHC copper box, two factors are considered when normalizing to simulated decays in the PMT. One is the mass of PMT and copper box. The other is the activity of each isotope in the PMT and OFHC copper. The activity of each isotope in the PMT is assumed to be ~0.01 ppb, which corresponds to 124 μBq/kg for ^{238}U, 40.5 μBq/kg for ^{232}Th and 2500 μBq/kg for ^{40}K, respectively. The activity of each isotope in the OFHC copper was obtained in Ref [7]. Table 2 shows the simulated decays in the copper box for each isotope.

Table 2. The number of decays simulated for each isotope in the copper box. Where, 8.2 in column 5 is the mass ratio between the copper box (1656 g) and PMT (202 g).

Isotopes	Activity in the PMT (μBq/kg)	Activity in the OFHC copper (μBq/kg)	Simulated decays in the PMT	Simulated decays in the copper box
^{238}U	124	16	10^7	$10^7 \times \frac{16}{124} \times 8.2 = 4.058 \times 10^7$
^{232}Th	40.5	19	10^7	$10^7 \times \frac{19}{40.5} \times 8.2 = 3.85 \times 10^7$
^{40}K	2500	88	10^7	$10^7 \times \frac{88}{2500} \times 8.2 = 2.886 \times 10^6$

3. Monte Carlo analysis

3.1. Detection Efficiency

The basic definition of detection efficiency for photon can be expressed as follows [8]:

$$\varepsilon = \frac{\textit{total number of detected photons in the full-energy peak}}{\textit{total number of photons emitted by the source}} \tag{1}$$

Where, *total number of photons emitted by the source* is the product of the total decays (*10 million/number of radionuclide in the decay chain of* ^{238}U, ^{232}Th *or* ^{40}K) and the branching ratio for the photon of interest. The Compton background continuum was taken into account when counting the *total number of detected photons in the full-energy peak* since Compton events are the primary source of background counts under the full-energy peak [8].

Two types of detector efficiency have been studied in simulation I: (1) the five individual detector efficiency (ε_{D0}, ε_{D1}, ε_{D2}, ε_{D3}, and ε_{D4}), and (2) the total detector efficiency from the sum of energy deposition in five individual detectors (ε_{total}). Note that the detection threshold was assumed to be 10 keV for the well-shaped Ge detector array.

Journal of Physics: Conference Series **606** (2015) 012019 doi:10.1088/1742-6596/606/1/012019

Table 3, table 4 and table 5 show the simulation results of these three types of detector efficiencies for each selected gamma ray of interest from the decay chain of ^{238}U, ^{232}Th and ^{40}K, respectively.

Table 3. Detector efficiency for each selected gamma ray of interest from ^{238}U decay chain.

Radionuclide	E_γ (keV)	Branching ratio [9]	ε_{D0} (%)	ε_{D1} (%)	ε_{D2} (%)	ε_{D3} (%)	ε_{D4} (%)	ε_{total} (%)
^{234}Th	92	0.025	5.4	1.58	1.55	1.40	1.64	11.4
^{234}Pa	1001	0.0056	11.3	2.54	2.44	2.57	2.52	24.3
^{226}Ra	186	0.035	15.2	4.22	4.24	4.08	4.39	28.6
^{214}Pb	352	0.342	12.0	3.15	3.00	3.09	3.13	26.6
	295	0.177	12.7	3.36	3.24	3.25	3.21	27.9
^{214}Bi	2204	0.046	4.11	1.17	1.20	1.10	1.20	11.3
	1847	0.002	5.60	1.31	1.35	1.41	1.26	19.5
	1764	0.147	4.53	1.25	1.29	1.19	1.21	12.2
	1120	0.141	4.08	1.47	1.41	1.49	1.51	6.46

Table 4. Detector efficiency for each selected gamma ray of interest from ^{232}Th decay chain.

Radionuclide	E_γ (keV)	Branching ratio [10]	ε_{D0} (%)	ε_{D1} (%)	ε_{D2} (%)	ε_{D3} (%)	ε_{D4} (%)	ε_{total} (%)
^{228}Ac	969	0.162	5.56	1.61	1.64	1.60	1.57	13.4
	911	0.266	5.87	1.64	1.63	1.59	1.61	13.9
	338	0.113	11.7	3.20	3.25	3.23	3.20	24.0
^{228}Th	84.4	0.013	4.83	0.982	0.99	1.05	1.02	9.08
^{224}Ra	241	0.004	13.6	3.36	3.36	3.12	3.45	28.2
^{212}Pb	239	0.436	13.2	3.39	3.35	3.37	3.39	28.1
^{212}Bi	727	0.067	7.03	2.13	2.04	2.08	1.98	15.5
^{208}Ti	583	0.304	5.44	1.96	1.92	1.91	1.95	8.22

Table 5. Detector efficiency for each selected gamma ray of interest from ^{232}Th decay chain.

Radionuclide	E_γ (keV)	Branching ratio	ε_{D0} (%)	ε_{D1} (%)	ε_{D2} (%)	ε_{D3} (%)	ε_{D4} (%)	ε_{total} (%)
^{40}K	1461	0.105	4.63	1.25	1.23	1.26	1.25	12.10

The results of simulation I are summarized in tables 3 - 5, which show that the Ge planar detector, D0 (at the bottom of the well), is most sensitive to the signal from the PMT, indicating that the individual detector efficiency has a strong position dependence. This property can be used to track gamma rays from the PMT and other detector components for a better understanding of the backgrounds. Moreover, due to the well-shaped geometry, the total detector efficiency of the proposed detector array for low-energy γ-rays has a factor of ~2 improvement when comparing to that of the single planar Ge detector with best detector efficiency.

3.2. Minimum detectable activity (MDA)

In addition to the detector efficiency, we have also studied another detector performance, Minimum Detectable Activity (MDA) in simulation II.

The Minimum Detectable Activity (MDA) is defined as the smallest amount of radioactivity that can be distinguished from a blank sample. When the count rate of a sample is roughly the same as the count rate of the background, the MDA becomes significant in low-level counting. The MDA for γ-ray measurement system can be calculated according to the equation [11] below:

$$MDA = \frac{\alpha\sqrt{N_b}}{r_b \Delta T \varepsilon M} \tag{2}$$

Where, MDA is in Bq/kg or Bq/PMT, α is a constant that equals to 1.64 at 90% confidence level (C.L.), N_b is the number of background counts in the region of interest (ROI) for a γ ray from a target redionuclide, r_b is the branching ratio of the selected γ ray from the target radionuclide, ΔT is the counting time in seconds, ε is the full absorption efficiency for the selected γ ray, and M is the mass of the counting sample, which is the PMT (201.47 g) in our work.

r_b and ε can be directly obtained from simulation I (table 3 ~ 5). ΔT can be calculated from the given total simulated decays and activity in OFHC copper for each isotope, which are $2.66 \times 10^{10}s$ (4.4×10^4 weeks), $1.11 \times 10^{11}s$ (1.84×10^5 weeks), and $1.98 \times 10^{10}s$(3.27×10^4 weeks) for ^{238}U, ^{232}Th and ^{40}K, respectively. Hence, the only parameter needs to be determined in Eq. 2 is N_b, the number of backgrounds counts in the ROI, to calculate the MDA for each selected gamma ray from the decay chain of ^{238}U, ^{232}Th and ^{40}K. The energy resolution for each selected gamma ray needs to be taken into account in order to get the corresponding background counts, N_b, from simulation II. 12 gamma rays and their corresponding Full Width at Half Maximum (HWHM) from Ref. [12] were used to fit the energy resolution. The best fit function is presented in Eq. 3:

$$FWHM = (1.423 \pm 0.019) + (6.168 \pm 0.639) \times 10^{-5} \cdot E_r - (6.475 \pm 4.391) \times 10^{-8} \cdot E_r^2 \tag{3}$$

Where, E_r is the energy of the selected γ ray. The energy resolution in terms of FWHM as a function of the γ ray energy is plotted in Fig. 2.

With Eq. 3, the $FWHM$ and thus the MDA of each selected gamma ray can then be calculated.

Table 6, table 7 and table 8 show the simulation results of $FWHM$ and MDA for each selected gamma rays from ^{238}U, ^{232}Th and ^{40}K, respectively. From the MDA results, the detector sensitivity was not good enough with the counting time constrained to one week.

4. Conclusion

To have a good understanding of the radiogenic backgrounds, especially the radiogenic gamma rays from the PMTs that are used in the LUX and XENON100 detectors, we have designed a Ge-based well detector and a R11410MOD PMT in a Geant4-based Monte Carlo simulation. Two independent simulations were conducted for studying the detector response to the signal from the PMT and the detector response to the background from the copper boxes that are around the Ge planar detectors. In these two simulations, the detector performance including three types of detector efficiency, energy resolution and the Minimum Detectable Activity (MDA)

Journal of Physics: Conference Series **606** (2015) 012019 doi:10.1088/1742-6596/606/1/012019

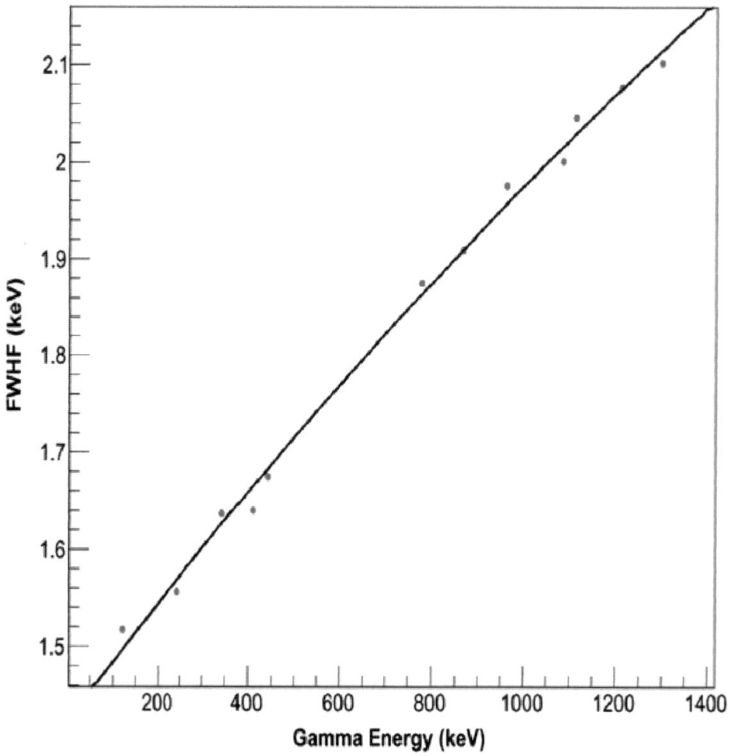

Figure 2. Energy resolution (FWHM) vesus the energy of gamma rays.

Table 6. $FWHM$ and MDA for each selected gamma ray from ^{238}U decay chain.

Raduinuclide	E_γ (keV)	$FWHM$ (keV)	ΔT (s)	N_b	MDA (mBq/kg)	MDA (mBq/PMT)
^{234}Th	92	1.48	2.66E+10	14189	1.28E-02	6.32E-02
^{234}Pa	1001	1.98	2.66E+10	1502	8.69E-03	4.30E-02
^{226}Ra	186	1.54	2.66E+10	14020	3.61E-03	1.79E-02
^{214}Pb	352	2.46	2.66E+10	69422	8.84E-04	4.38E-03
	295	2.34	2.66E+10	41708	1.26E-03	6.25E-03
^{214}Bi	2204	1.6	2.66E+10	2823	3.12E-03	1.54E-02
	1847	1.64	2.66E+10	1813	3.33E-02	1.65E-01
	1764	2.04	2.66E+10	9719	1.68E-03	8.31E-03
	1120	2.32	2.66E+10	9145	3.20E-03	1.59E-02

have been studied. According to the simulation results, we can conclude that this proposed germanium well detector has the capability of being sensitive to the PMTs with radioactivity

Table 7. $FWHM$ and MDA for each selected gamma ray from ^{232}Th decay chain.

Raduinuclide	E_γ (keV)	$FWHM$ (keV)	ΔT (s)	N_b	MDA (mBq/kg)	MDA (mBq/PMT)
^{228}Ac	969	1.96	1.11E+11	67922	5.69E-02	2.81E-01
	911	1.94	1.11E+11	114699	3.18E-01	1.57E+00
	338	1.62	1.11E+11	109630	2.88E-02	1.43E-01
^{228}Th	84.4	1.48	1.11E+11	25738	3.78E-03	1.87E-02
^{224}Ra	241	1.56	1.11E+11	53904	3.40E-03	1.68E-02
^{212}Pb	239	1.56	1.11E+11	8374	5.17E-03	2.56E-02
^{212}Bi	727	1.84	1.11E+11	38301	4.61E-01	2.28E+00
^{208}Ti	583	1.76	1.11E+11	129893	2.18E-02	1.08E-01

Table 8. $FWHM$ and MDA for the selected gamma ray from ^{40}K decay chain.

Raduinuclide	E_γ (keV)	$FWHM$ (keV)	ΔT (s)	N_b	MDA (mBq/kg)	MDA (mBq/PMT)
^{40}K	1461	2.18	1.98E+10	29794	2.34E-02	1.16E-01

around 0.01 ppb.

References

[1] Akerib D S *et al.* (LUX Collarboration) 2013 *NIM A*, Volume 703, p. 1-6. arXiv:1205.2272v2.
[2] Aprile E *et al.* (XENON100 Collaboration) 2012 *Phys. Rev. Lett.* **109** 181301 arXiv:1207.5988.
[3] Akerib D S *et al.* (LUX Collaboration) 2013 *Phys. Rev. Lett.* **112** 091303 arXiv:1310.8214.
[4] LZ Collaboration, http://lz.lbl.gov/.
[5] Aprile E *et al.*(XENON1T Collaboration) 2012, arXiv:1206.6288.
[6] Refer to: http://www-eng.lbl.gov/ shuman/NEXT/MATERIALS&COMPONENTS/PMT/PMT_R11410MOD_2010-0323.pdf.
[7] Heusser G *et al.*, Radioactivity in the Environment, DOI: 10.1016/S1569-4860 (05) 08039-3.
[8] Refer to http://www.lanl.gov/orgs/n/n1/panda/00326398.pdf.
[9] Refer to http://hepwww.rl.ac.uk/ukdmc/radioactivity/U238_chain/U238_chain.html.
[10] Refer to http://hepwww.rl.ac.uk/ukdmc/radioactivity/Th_chain/Th_chain.html.
[11] Currie LA 1968 *Analytical Chemistry* 40 (**3**) 587-593.
[12] Abhigyan A, Resolution and efficiency of high purity germanium, Refer to: http://www.academia.edu/4191151/Resolution_and_Efficiency_of_High_Purity_Germanium_Detector.

AUTHOR INDEX